T0239191

SpringerBriefs in Energy

SpringerBriefs in Energy presents concise summaries of cutting-edge research and practical applications in all aspects of Energy. Featuring compact volumes of 50 to 125 pages, the series covers a range of content from professional to academic. Typical topics might include:

- A snapshot of a hot or emerging topic
- A contextual literature review
- A timely report of state-of-the art analytical techniques
- An in-depth case study
- A presentation of core concepts that students must understand in order to make independent contributions.

Briefs allow authors to present their ideas and readers to absorb them with minimal time investment.

Briefs will be published as part of Springer's eBook collection, with millions of users worldwide. In addition, Briefs will be available for individual print and electronic purchase. Briefs are characterized by fast, global electronic dissemination, standard publishing contracts, easy-to-use manuscript preparation and formatting guidelines, and expedited production schedules. We aim for publication 8–12 weeks after acceptance.

Both solicited and unsolicited manuscripts are considered for publication in this series. Briefs can also arise from the scale up of a planned chapter. Instead of simply contributing to an edited volume, the author gets an authored book with the space necessary to provide more data, fundamentals and background on the subject, methodology, future outlook, etc.

SpringerBriefs in Energy contains a distinct subseries focusing on Energy Analysis and edited by Charles Hall, State University of New York. Books for this subseries will emphasize quantitative accounting of energy use and availability, including the potential and limitations of new technologies in terms of energy returned on energy invested.

More information about this series at https://link.springer.com/bookseries/8903

Futoshi Matsumoto · Takao Gunji

Water in Lithium-Ion Batteries

 Springer

Futoshi Matsumoto
Department of Materials and Life
Chemistry
Kanagawa University
Yokohama, Japan

Takao Gunji
Department of Materials and Life
Chemistry
Kanagawa University
Yokohama, Japan

ISSN 2191-5520   ISSN 2191-5539   (electronic)
SpringerBriefs in Energy
ISBN 978-981-16-8785-3   ISBN 978-981-16-8786-0   (eBook)
https://doi.org/10.1007/978-981-16-8786-0

This Springer imprint is published by the registered company Springer Nature Singapore Pte Ltd.
The registered company address is: 152 Beach Road, #21-01/04 Gateway East, Singapore 189721,
Singapore

Preface

Water (H_2O) in lithium-ion batteries (LIBs), which are constructed with anodes, cathodes and organic electrolytes that contain lithium salts, can degrade the cell performance and seriously damage the materials. However, because a small amount of H_2O in cells contributes to the formation of a solid electrolyte interphase (SEI), the complete removal of H_2O from cells lowers battery performance and increases the expense of H_2O removal from the battery materials. The optimal concentration of H_2O for each battery material has been determined, and these concentrations are maintained with appropriate removal methods and H_2O scavengers that were recently developed to establish both high performance and low cost. More recently, to achieve both the safety and low cost of LIBs, the development of anode and cathode preparations by aqueous processes and aqueous LIBs in which aqueous electrolytes containing lithium salts are used as electrolytes has progressed. In this review, information on the H_2O content in LIBs, the reactivity of anodes, cathodes and electrolytes with water and the processes underlying H_2O resistance in LIB materials is reviewed from the perspective of H_2O concentration and LIB stability. The goal of this review is to provide appropriate information concerning the amount of H_2O needed in cells to achieve stable and high cell performance.

Yokohama, Japan

Futoshi Matsumoto
Takao Gunji

Contents

Chapter 1
Introduction

Abstract In this chapter, history of lithium-ion batteries (LIBs) and their problems which should be resolved in future are overviewed. In addition, the problem caused by water which is main part of this book is raised with the reaction occurred with water in LIBs. Finally, the purpose and the content of this book is explained.

Keywords Water-resistant properties · Moisture · Lithium salt · Hydrogen fluoride · Solid-state electrolytes · Rechargeable aqueous batteries

In the thirty years since Sony successfully commercialized LIBs in 1991, they have achieved remarkable progress in the area of portable devices, in which these batteries are used as power sources [1–4]. The progress of electronic devices and LIBs is observed through the establishment of wearable devices [5]. Because of the high energy density of LIBs, the development of automobiles using LIBs as power sources has also arrived at the level of automobile commercialization, with comparable performance to engine-driven automobiles [6–8]. LIBs have received attention as load-leveling battery energy storage devices [9, 10]. Improvements in the battery performance of LIBs will be increasingly needed to improve our daily lives. Together with the improvement of battery performance, the safety and robustness of LIBs as well as the reduction in the cost of LIBs have become critical issues [11, 12]. Low-cost LIBs produced by environmentally friendly processes would certainly be desirable in the next generation of LIBs [13–15]. In addition, new Navy and undersea applications in which LIBs contact water (H_2O) have received attention because of their excellent energy density, reliability and life in commercial applications of LIBs when compared with other power sources [16].

Among the currently commercially available LIBs, invasion of H_2O into batteries causes the degradation of battery performance [17]. H_2O causes problems in conventional LIBs [18]. At the production level of LIBs, contact of cathode materials with moisture causes serious problems in the reaction of the cathode particle surface with moisture, resulting in degradation of the cathode performance [19, 20]. H_2O can also react with electrolytes for LIBs, especially with lithium hexafluorophosphate ($LiPF_6$), to decompose the salt and decrease the concentration of lithium salt in the electrolytes [21]. Hydrogen fluoride (HF) (Eqs. 1.1 and 1.2) created by the decomposition of $LiPF_6$ in the presence of trace amounts of H_2O in LIB cells reacts with active materials of anodes and cathodes or the protective solid electrolyte interphase (SEI) on the anode, leading to the degradation of the anode and cathode performance [22].

$$LiPF_6 \rightarrow LiF + PF_5 \tag{1.1}$$

$$PF_5 + H_2O \rightarrow 2HF + POF_3 \tag{1.2}$$

Recently, from the perspective of the environmental feasibility of LIBs, the development of a water process in the fabrication of anodes and cathodes with water-soluble and aqueous polymer (water-based polymer) binders has attracted much attention [23–25]. In addition, from the perspective of the safety and lower cost of LIBs, rechargeable aqueous batteries (RABs), in which conventional lithium transition metal oxide materials and graphite coated with gel polymer membranes or lithium superionic conductors are used for cathodes and anodes and aqueous electrolytes that contain lithium salts function as electrolytes, are in high demand as green energy storage devices [26, 27]. Therefore, the relationship between H_2O and materials in LIBs should be examined carefully, and water-resistant properties should be properly provided in the materials. Some solid-state electrolytes that have rapidly received increased attention in recent years also react with H_2O and lower the performance of solid-state electrolytes, and the resistance of the interfaces between anodes/cathodes and solid-state electrolytes has seriously increased. A schematic description of the reactions and processes related to H_2O in LIBs, which are discussed in this book, is summarized in Fig. 1.1.

In this review book, first, the disadvantage of H_2O in LIBs will be summarized, and then, the degradation mechanism of cathode and anode materials and electrolytes caused by contact with H_2O will be discussed. The protective method against performance degradation caused by H_2O will be reviewed. Finally, the water process in the fabrication of anodes and cathodes and the systems of RABs and all-solid-state LIBs, as well as their advantages and disadvantages, will be explained. Throughout this review, questions concerning the amount of H_2O that affects the properties of materials in LIBs and battery performance will be addressed and summarized as much as possible.

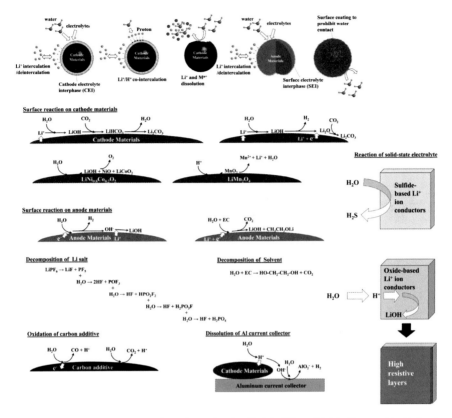

Fig. 1.1 Schematic description of reactions related to H_2O in LIBs, which is discussed in this book

References

1. Nishi Y (2001) J Power Sources 100:101
2. Goodenough JB, Kim Y (2010) Chem Mater 22:587
3. Li M, Lu J, Chen Z, Amine K (2018) Adv Mater 30:1800561
4. Liang Y, Zhao CZ, Yuan H, Chen Y, Zhang W, Huang JQ, Yu D, Liu Y, Titirici MM, Chueh YL, Yu H, Zhang Q (2019) InfoMat 1:6
5. Kim SH, Choi KH, Cho SJ, Choi S, Park S, Lee SY (2015) Nano Lett 15:5168
6. Wagner FT, Lakshmanan B, Mathias MF (2020) J Phys Chem Lett 1:2204
7. Bresser D, Hosoi K, Howell D, Li H, Zeisel H, Amine K, Passerini S (2018) J Power Sources 382:176
8. Marinaro M, Bresser D, Beyer E, Faguy P, Hosoi K, Li H, Sakovica J, Amine K, Wohlfahrt-Mehrens M, Passerini S (2020) J Power Sources 459: 228073
9. Yang Y, Li H, Aichhorn A, Zheng J, Greenleaf M (2014) IEEE Trans Smart Grid 5:982
10. Yoo HD, Markevich E, Salitra G, Sharon D, Aurbach D (2014) Mater Today 17:110
11. Wu X, Song K, Zhang X, Hu N, Li L, Li W, Zhang L, Zhang H (2019) Front Energy Res 7:65
12. Wang Q, Ping P, Zhao X, Chu G, Sun J, Chen C (2012) J Power Sources 208:210
13. Wang J, Nie P, Ding B, Dong S, Hao X, Dou H, Zhang X (2017) J Mater Chem A 5:2411
14. Schmuch R, Wagner R, Hörpel G, Placke T, Winter M (2018) Nat Energy 3:267
15. Dunn JB, Gaines L, Kelly JC, James C, Gallagher KG (2015) Energy Environ Sci 8:158

16. Gitzendanner R, Puglia F, Martin C, Carmen D, Jones E, Eaves S (2004) J Power Sources 136:416
17. Kawamura T, Okada S, Yamaki J (2006) J Power Sources 156:547
18. Zheng LQ, Li SJ, Lin HJ, Miao YY, Zhu L, Zhang ZJ (2014) Russian J Electrochem 50:904
19. Huang B, Qian K, Liu Y, Liu D, Zhou K, Kang F, Li B (2019) ACS Sustainable Chem Eng 7:7378
20. Liu HS, Zhang ZR, Gong ZL, Yang Y (2004) Electrochem Solid-State Lett 7:A190
21. Li W, Lucht BL (2017) Electrochem Solid-State Lett 10:A115
22. Wu Y, Jiang C, Wan C, Tsuchida E (2000) Electrochem Commun 2:626
23. Chen Z, Kim GT, Chao D, Loeffler N, Copley M, Lin J, Shen Z, Passerini S (2017) J Power Sources 372:180
24. Du Z, Rollag KM, Li J, An SJ, Wood M, Sheng Y, Mukherjee PP, Daniel C, Wood DL III (2017) J Power Sources 354:200
25. Hwa Y, Frischmann PD, Helms BA, Cairns EJ (2018) Chem Mater 30:685
26. Chang Z, Li C, Wang Y, Chen B, Fu L, Zhu Y, Zhang L, Wu Y, Huang W (2016) Sci Rep 6:28421
27. Alias N, Mohamad AA (2015) J Power Sources 274:237

Chapter 2
Water Content in LIBs

Abstract The amount of water which was carried in LIBs in various steps of the manufacturing process of LIBs is summarized. The reaction mechanism of decomposition of cathode and anode materials and electrolytes and analysis method for water concentration are explained. Several water removal methods and their efficiency are discussed.

Keywords H_2O content · H_2O uptake · H_2O removal · Casting solvents · Decomposition mechanism

2.1 Drying Processes of Cell Components

The residual H_2O present in LIB cells has been widely regarded as a detrimental factor for the performance of LIBs. Therefore, the amount of H_2O in cells should be controlled during the manufacturing process of LIBs. The moisture content in the dry rooms in which battery packs are fabricated should remain below 100 parts per million (ppm) [1]. Dry rooms are designed to efficiently control the moisture in the room to decrease the H_2O content to below 100 ppm [1, 2]. Cathode and anode materials, separators and electrolyte solutions are very hygroscopic. Therefore, H_2O is taken up by cells from the anode and cathode materials, separators and electrolyte solutions containing H_2O. All components of the LIBs are dried thoroughly before preparing cathodes and anodes and assembling the LIB cells. Many papers examining the drying behavior of components have been reported. Stich et al. investigated the drying behavior and H_2O uptake of a variety of commonly used electrode materials (graphite, $LiFePO_4$ (LFP), $LiMn_2O_4$, $LiCoO_2$, $Li(NiCoMn)O_2$) and separators (polyolefin, glass fiber) [3]. To categorize the strength of H_2O bonding to the materials, three segments for the heating process were applied in the H_2O emission experiments. The temperature and heating time were optimized for each component. From the obtained results (Fig. 2.1), it was found that the H_2O contents of the graphite anode and glass fiber separator were highest among the tested materials and that the investigated cathode materials showed a wide variation in drying behavior and a low H_2O content in the case of $LiCoO_2$, moderate contents for the cases of $LiMn_2O_4$ and $Li(NiCoMn)O_2$ and a high content in the case of LFP [3].

F. Matsumoto and T. Gunji, *Water in Lithium-Ion Batteries*, SpringerBriefs in Energy, https://doi.org/10.1007/978-981-16-8786-0_2

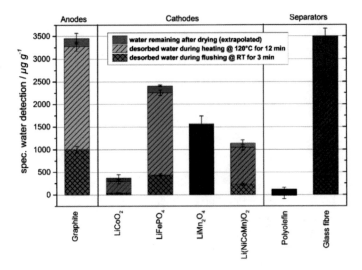

Fig. 2.1 Comparison of specific H_2O contents of different anodes, cathodes and separators detected at room temperature (cross-hatched), during heating at 120 °C (hatched) and remaining after the experiment runtime, as determined by extrapolation (solid) [3]. Reprinted from J. Power Sources, 364, Stich, M., Pandey, N., Bund, A.: Drying and moisture resorption behavior of various electrode materials and separators for lithium-ion batteries, 84–91, Copyright (2017), with permission from Elsevier

When cathodes and anodes are prepared for LIBs, slurries that are prepared by mixing cathode or anode materials, conductive additives and binders with solvents are cast on current collectors, and then casting solvents are evaporated from the slurries to form the cathode and anode layers by heating the slurries on the current collectors. Water uptake of cathode and anode electrodes may also occur from the casting solvents because these solvents may contain H_2O. The conditions under which the heating process evaporates the casting solvent should be strictly optimized so that a small amount of H_2O is not left. Acetone has been used as a model casting solvent, not N-methylpyrrolidone (NMP). The acetone wetness and the length of exposure of cathode and anode materials to acetone did not contribute to H_2O uptake by cathode and anode materials [4]. Huttner et al. investigated the influence of the post-drying process on H_2O content and battery performance [5]. The post-drying process was applied to the cell components, cathodes, anodes and separators to avoid remoistening the components before assembling LIB cells. Sufficiently drying the cell components caused low battery performance, and medium post-drying (the H_2O content was 326 ppm across all electrodes) resulted in good battery performance. They discussed the reason why sufficient drying of cathodes or anodes caused degradation of battery performance. It was also found that the electrical properties of cathodes and anodes declined. The damage of electric percolation pathways with changes in cathode and anode layers on current collectors by sufficient drying was considered one of the causes of battery performance degradation.

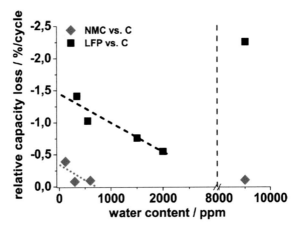

Fig. 2.2 Relative capacity loss for various cell compositions depending on the H_2O content within the cathode [6]. Reprinted from J. Ceram. Sci. Tech., 4, Langklotz, U.; Schneider, M.; Michaelis, A. Water uptake of tape-cast cathodes for lithium ion batteries, 69–76, Copyright (2013), with permission from Göller Verlag GmbH

After preparation of the cathodes, during storage of the cathodes, the kinetics of H_2O uptake and release to/from the cathodes were also examined with LFP and $Li(Mn_{0.37}Co_{0.35}Ni_{0.35})O_{2.07}$ cathode materials [6]. H_2O uptake on the surface of cathode materials follows the adsorption model of Brunauer, Emmett and Teller (BET). H_2O uptake to cathode materials reaches a steady state after one hour of exposure. The steady-state values of H_2O uptake were approximately 2000 and 700 ppm for LFP and $Li(Mn_{0.37}Co_{0.35}Ni_{0.35})O_{2.07}$, respectively. To release H_2O from the cathodes to less than 100 ppm, drying temperatures of 200 and 150 °C for LFP and $Li(Mn_{0.37}Co_{0.35}Ni_{0.35})O_{2.07}$, respectively, were needed for 15 h. The performance of the LFP and $Li(Mn_{0.37}Co_{0.35}Ni_{0.35})O_{2.07}$ cathodes, which contained various H_2O contents, was examined by preparing cells with graphite anodes. Relative capacity losses (%/cycle), which were calculated based on the results of charging/discharge cycles up to 30 cycles, decreased with increasing amounts of H_2O in the cathodes in a range of H_2O contents less than 2000 ppm. The LFP cathode was sensitive to a high H_2O content (Fig. 2.2). The high sensitivity of LFP was considered to be due to the carbon coating on LFP particles. The H_2O content in a cell could be connected to the composition and insulating properties of the SEI. The relationship between the formation of SEI and H_2O content will be discussed later.

2.2 Amount of Water in a Cell and Removal of Water from a Cell

Dahn et al. examined the influence of the intentional addition of H_2O to a cell on the battery performance of $LiCoO_2$/graphite and $Li[Ni_{0.42}Mn_{0.42}Co_{0.16}]O_2$/graphite cells [7] and $LiCoO_2/Li_4Ti_5O_{12}$ (LTO) [8] with 1 M $LiPF_6$ in ethylene carbonate (EC):ethyl methyl carbonate (EMC) (3:7 by wt.). Intentional addition of H_2O up to 1000 ppm to the electrolyte did not affect the battery performance. The authors suggested that the results indicated the possibility of relaxation of H_2O content specifications in electrolyte lading to one avenue for cost reduction of LIBs.

All the above-mentioned results described for H_2O content were obtained by Karl Fisher titration [4–8]. However, Karl Fisher titration cannot be applied for real-time monitoring of H_2O content in LIB cells. New *operando* fluorescence (FL) spectroscopy with nanosized coordination polymers was proposed to monitor trace H_2O during electrochemical cycles [9]. Nanosized coordination polymers were added to electrolytes and provided a distinguishable turn-on FL response toward H_2O with a quantifiable detection range from 0 to 1.2% v/v. With the monitoring system, trace H_2O was indeed generated during the first discharge process of graphite anodes based on the observations of increases in the fluorescence intensity over time along with the gradual formation of H_2O. The amount of H_2O generated was calculated to be approximately 0.18% in 2 mL of electrolyte. The mechanism of H_2O formation during the first discharge process was not discussed in the paper.

Although the H_2O content should be low in LIB cells, H_2O removal requires time and cost. In addition, H_2O is formed during the charging/discharging cycles [10, 11]. For example, Yu et al. proposed decomposition mechanisms for EC solvent and $LiPF_6$ salt upon oxidation in the presence of transition metal oxide. H_2O is formed during the reaction (Fig. 2.3) [10].

Jung et al. proposed a mechanism for the oxidation of EC with reactive oxygen species (e.g., singlet oxygen) released from the $LiNi_xMn_yCo_zO_2$ (NMC) structure that yielded carbon dioxide (CO_2), carbon monoxide (CO) and H_2O [11]. Functions for continuous removal of H_2O should be equipped with LIBs as well as a sufficient drying process for cell components before cell assembly. To continuously remove H_2O from the cells, the development of additives that are inherent in packed cells

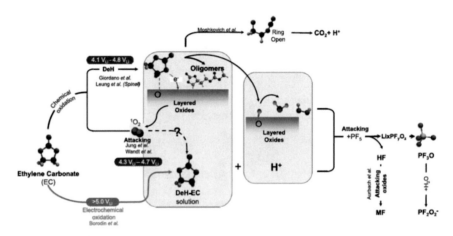

Fig. 2.3 Proposed decomposition mechanisms of EC solvent and $LiPF_6$ salt upon oxidation in the presence of the transition metal oxide, where EC molecules dissociate on the oxide surface, forming dehydrogenated species and surface protic species or H_2O. Reprinted with permission from Ref. [10]. Copyright 2018 American Chemical Society

has been considered. Primary additives are molecules that can scavenge H_2O or products formed by reacting Li salts such as $LiPF_6$ with H_2O [12]. For example, 0.1 wt.% N'N-1,4-phenylenedimaleimide (MI) was added to a mixture of 1 M $LiPF_6$, EC, propylene carbonate (PC) and diethylene carbonate (DEC) (3:2:5 by volume) containing 100 ppm H_2O [13]. It was mentioned in the paper that MI reacts with H_2O and $LiPF_6$ to form a temperature-stable three-dimensional nanotunnel in the SEI on graphite anodes. Dimethyl acetamide (DMAc) also functions as a scavenger in 1 M $LiPF_6$, EC, DEC and dimethyl carbonate (DMC) (1:1:1: by volume) containing 0.5 wt.% H_2O and inhibits the generation of HF and other decomposition products of $LiPF_6$ [14]. Copper-1,3,5-benzenetricarboxylic acid (CuBTC) has also been used as a scavenger for H_2O. CuBTC forms metal–organic frameworks (MOFs) featuring high surface areas, abundant pores and diversified metal sites. The unique micropores efficiently capture H_2O. CuBTC MOF powder was mixed with a polymer binder to prepare a CuBTC MOF film, which was used as a separator in a LIB cell [15]. The cell constructed with the CuBTC MOF separator, high-voltage cathodes ($LiNi_{0.5}Mn_{1.5}O_4$, $LiNi_{0.8}Co_{0.1}Mn_{0.1}O_2$), Li metal anode and electrolytes containing 200 ppm H_2O exhibited the same behavior as the one obtained with a cell prepared with high-voltage cathodes ($LiNi_{0.5}Mn_{1.5}O_4$, $LiNi_{0.8}Co_{0.1}Mn_{0.1}O_2$), Li metal anode, conventional separator and pre-dried electrolytes. Dahn also examined the scavengers of H_2O in LFP/graphite cells. They reported that cells containing electrolyte additives were generally not affected by $\sim$500 ppm H_2O in the cell [16]. The scavengers that capture the decomposition products formed by reaction with $LiPF_6$ and H_2O have been reviewed in detail by Choi [12].

The data on the amount of H_2O taken up a cell in LIBs are fragmentary, and no systematic study of the effect of H_2O on LIBs has been conducted. Many markers of components in LIBs might individually collect data on their own materials. The data might be negative for each material maker. However, we think that to reduce the cost of LIBs and to reduce the development time for LIBs, a maximum permissible level of H_2O content at which battery performance is not affected should be systematized.

References

1. Nishi Y (2001) J Power Sources 100:101
2. Goodenough JB, Kim Y (2010) Chem Mater 22:587
3. Li M, Lu J, Chen Z, Amine K (2018) Adv Mater 30:1800561
4. Liang Y, Zhao CZ, Yuan H, Chen Y, Zhang W, Huang JQ, Yu D, Liu Y, Titirici MM, Chueh YL, Yu H, Zhang QA (2019) InfoMat 1:6
5. Kim SH, Choi KH, Cho SJ, Choi S, Park S, Lee SY (2015) Nano Lett 15:5168
6. Wagner FT, Lakshmanan B, Mathias MF (2010) J Phys Chem Lett 1:2204
7. Bresser D, Hosoi K, Howell D, Li H, Zeisel H, Amine K, Passerini S (2018) J Power Sources 382:176
8. Marinaro M, Bresser D, Beyer E, Faguy P, Hosoi K, Li H, Sakovica J, Amine K, Wohlfahrt-Mehrens M, Passerini S (2020) J Power Sources 459: 228073
9. Yang Y, Li H, Aichhorn A, Zheng J, Greenleaf M (2014) IEEE Trans Smart Grid 5:982
10. Yoo HD, Markevich E, Salitra G, Sharon D, Aurbach D (2014) Mater Today 17:110

11. Wu X, Song K, Zhang X, Hu N, Li L, Li W, Zhang L, Zhang H (2019) Front Energy Res 7:65
12. Wang Q, Ping P, Zhao X, Chu G, Sun J, Chen C (2012) J Power Sources 208:210
13. Wang J, Nie P, Ding B, Dong S, Hao X, Dou H, Zhang X (2017) J Mater Chem A 5:2411
14. Schmuch R, Wagner R, Hörpel G, Placke T, Winter M (2018) Nat Energy 3:267
15. Dunn JB, Gaines L, Kelly JC, James C, Gallagher KG (2015) Energy Environ Sci 8:158
16. Gitzendanner R, Puglia F, Martin C, Carmen D, Jones E, Eaves S (2004) J Power Sources 136:416

Chapter 3
Water-Sensitivity and Waterproof Features of Cell Components

Abstract Water-sensitivity of LIB components, anodes, cathodes, separators and electrolytes, is summarized with the mechanism of reaction with H_2O. Especially, the results of the aging of the materials upon exposure to humidity are collected from many papers and summarized in tables to understand which materials have waterproof property. The protection methods to give the materials waterproof property are also explained.

Keywords Capacity fading · Delithiation · Ion exchange · Washing · Gas formation · Li metal

3.1 Sensitivity of Anode Materials to Humidity and Their Protection Against Humidity

As mentioned above [1], graphite for an anode material is hygroscopic when compared with cathode materials. Because H_2O adsorbed on the graphite surface influences the formation of SEI on anode surfaces, control of the amount of H_2O adsorbed on anode surfaces is a critical factor for maintaining battery performance. In addition, the H_2O contents in the styrene-butadiene rubber (SBR) binder, carboxymethyl cellulose (CMC) rheology additive and carbon black (CB) conductive additive that are mixed with anode materials to form anode layers must be added with extreme caution to control the H_2O content of anodes. Although CB is hydrophobic, SBR and CMC cause high moisture uptake of the entire anode structure despite its low mass fraction [2].

Graphite materials are sensitive to H_2O and show a fast fade in capacity under high humidity. To inhibit the capacity fading of anode materials, Holze et al. proposed a composite of natural graphite and copper (Cu) [3] or silver (Ag) [4, 5] for anodes. Metallic Cu and Ag were deposited on the graphite. The Cu and Ag deposits occupy active sites on the graphite surfaces for absorbing H_2O.

Li metal has received much attention because of its high capacity for anodes. However, owing to its high reactivity and uncontrolled dendrite growth, the Li metal anode is not yet in practical use. In addition, when Li metal contacts H_2O, the Li metal combusts with the production of a large amount of heat and H_2. The batteries eventually become dangerous. Various waterproof Li metal anodes have been proposed with various strategies and materials. Yang et al. proposed the coating of Li metal surfaces with wax and poly(ethylene oxide) for the Li metal surface to have waterproof properties. The coated Li metal surface exhibited battery performance stability after exposing the coated Li metal surface for 24 h in ambient air with a relative humidity of 70% and dipping it into H_2O (Fig. 3.1) [6]. Surprisingly, even after dipping it in H_2O, the coated Li metal anode exhibited the same discharge curve and capacity as the nontreated Li metal anode. Goodenough et al. designed a composite of graphite fluoride (GF)/lithium fluoride (LF)/Li metal as an anode that is stable in a humid atmosphere with a relative humidity (RH) of 20–35%. After exposing the GF/LF/Li anode to a humid atmosphere for 24 h, it exhibited a comparable specific capacity and cycling stability to fresh Li metal anodes that were not exposed to a humid atmosphere. Owing to the hydrophobic GF-LF layer, the GF-LF layer can prevent the Li metal surface from contacting H_2O in a humid atmosphere [7]. Like other examples, Li metal anodes encapsulated with cross-linked poly(vinylidenecohexafluoropropylene) (PVdF–HFP) [8] and densely packed Li_xM (M = Si, Sn, or Al) nanoparticles encapsulated by large graphene sheets [9] were proposed. Germanium (Ge) or germanium oxide (GeO_x) layers formed on Li metal, which have a high Young's modulus and are insoluble in H_2O, were applied as protective layers against H_2O.

Li–Ge alloy formed with the reaction and Li metal and Ge has a weak interfacial resistance and therefore works as a Li–ion conductor between liquid electrolyte and Li metal [10]. Methods for protecting Li metal anodes from contact with H_2O have evolved so that modified Li metal anodes can be applied to anodes in RABs, as mentioned later. H_2O, which can impede battery performance, is important for the inhibition of Li metal dendrite formation by controlling the amount of H_2O (25–50 ppm) in cells. A trace amount of HF formed by the reaction between $LiPF_6$ and H_2O is electrochemically reduced to form a uniform LiF-rich SEI layer on the Li metal surface. The LiF-rich SEI layer is formed uniformly on the Li metal surface, which inhibits the formation of Li dendrites [11]. The LiF-rich SEI layer provides faster Li^+ diffusion across the electrode–electrolyte interface through highly conductive LiF channels. The faster diffusion of Li^+ ions through the LiF layer can reduce the grain size of Li deposits [12]. Of course, the coulombic efficiency of lithium deposition/dissolution reactions is affected by the amount of H_2O in cells. The effect will be discussed later.

3.2 Sensitivity of Cathode Materials to Humidity and Their Protection Against Humidity

It is well known that cathode materials are damaged during storage under moisture. Therefore, care must be taken to dry the cathode materials sufficiently for the cathode

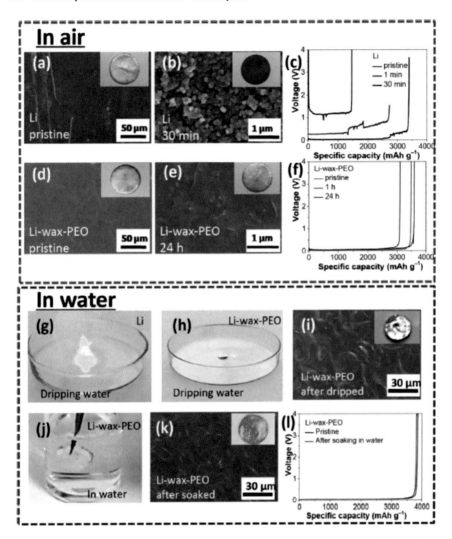

Fig. 3.1 Stability of Li-wax-PEO and Li in air and H_2O. **a, b** Morphologies of Li before and after exposure to air for 30 min (insert: the optical photographs), and **c** corresponding stripping curves; **d, e** morphologies of Li-wax-PEO before and after exposure to air for 24 h (insert: the optical photographs) and **f** corresponding stripping curves; optical photographs of **g** Li and **h** Li-wax-PEO when H_2O was dripped on the surface; **i** morphology of Li-wax-PEO after H_2O dripping (insert: the optical photographs); **j** optical photograph when Li-wax-PEO was soaked in H_2O and **k** morphology of Li-wax-PEO after soaking in H_2O (insert: the optical photographs); **l** stripping curves of Li-wax-PEO before and after soaking in H_2O; the relative humidity of the air was approximately 70% [6]. Reprinted from Sci. Bull., 64, Zhang, Y.; Lv, W.; Huang, Z.; Zhou, G.; Deng, Y.; Zhang, J.; Zhang, C.; Hao, B.; Qi, Q.; He, Y.-B.; Kang, F.; Yang, Q.-H.: An air-stable and waterproof lithium metal anode enabled by wax composite packaging, 910–917, Copyright (2019), with permission from Elsevier

materials not to degrade during storage. Li^+ ions are removed from the surfaces of cathode materials to react with H_2O and CO_2 in the atmosphere, and then lithium compounds such as lithium hydroxide (LiOH) and lithium carbonate (Li_2CO_3) are formed on the surfaces. The thicknesses of the LiOH and Li_2CO_3 layers are over a few nanometers. Almost all cathode materials exhibit the same behavior against moisture in the atmosphere during storage. The degree of the sensitivity of cathode materials against H_2O is different. Many researchers have examined the sensitivity against H_2O and reported the results. The results of examinations of the effect of moisture on the performance of cathode materials, such as the reduced capacity after contact with moisture and capacity retention for several charging/discharging cycles, are summarized in Table 3.1 [13–24]. Basically, cathode materials with a high Ni content exhibit a large degradation capacity after long-term storage under high humidity. For example, the dynamics of cathode–air interfacial reactions with H_2O on $LiNi_{1/3}Mn_{1/3}Co_{1/3}O_2$ (NMC333), $LiNi_{0.6}Mn_{0.2}Co_{0.2}O_2$(NMC622) and $LiNi_{0.8}Mn_{0.1}Co_{0.1}O_2$ (NMC811) were examined in situ (Fig. 3.2). On the NMC333 surface, the thickness of LiOH is undetectable, indicating inactivity against H_2O vapor. The NMC622 surfaces are readily passivated by an ultrathin LiOH layer. The growth of LiOH layers on NMC811 is much faster than that on NMC622. The rate of growth of LiOH layers is NMC811 > NMC622 > NMC333. The order of the LiOH formation rate is certainly in agreement with the amount of Ni content in NMC [25].

1. Capacity retention = (discharged capacity observed with air-exposed cathode materials)/(discharged capacity observed with cathode materials after storage without atmospheric exposure) × 100 (%)

The mechanism by which cathode materials are attacked by H_2O and damaged will be explained hereinafter. In examining the underlying mechanism, acid treatment of cathode materials, such as layered $LiCoO_2$ and Li_2MnO_3, spinel $LiMn_2O_4$ and LiV_3O_8 and LFP, is of central interest. Proton-for-lithium-ion exchange in lithiated-transition metal oxide cathode materials occurs during the acid treatment (Eq. 3.1).

$$x H^+(aq.) + Li_{x+y}M_zO_u \rightarrow xLi^+(aq.) + H_xLi_yM_zO_u \qquad (3.1)$$

The outward migration of Li^+ ions and the inward migration of protons into the oxide bulk to balance the charge from Li leaching into H_2O and H^+ intercalating into layered structures would drive equilibrium reactions (3.2) and (3.3) to the right side and facilitate reaction (3.4) [26].

$$Li^+ + H_2O \leftrightarrow H^+ + LiOH \qquad (3.2)$$

$$LiOH + CO_2 \leftrightarrow LiHCO_3 \qquad (3.3)$$

$$LiOH + LiHCO_3 \rightarrow H_2O + Li_2CO_3 \qquad (3.4)$$

Table 3.1 Summary of the aging of cathode materials upon exposure to humidity

Cathode materials	Synthesis condition of cathode materials	Humid atmosphere			Capacity fading					References
		relative humidity (%)	Temperature (°C)	Storage period	Cutoff voltage (V vs. Li/Li$^+$)	C-rate (C)	Cycle number	Capacity retention[1] (%)		
$LiNi_{1/3}Mn_{1/3}Co_{1/3}O_2$	A hydroxide route, using transition metal hydroxide and lithium carbonate as starting materials	60	20	0.5 h	3–4.3	1	30	90		[13]
$LiNi_{1/3}Mn_{1/3}Co_{1/3}O_2$	Amperex Technology Limited (ATL, China)	80	25	6 months	2.8–4.3	0.2	200	86		[14]
$LiNi_{0.6}Mn_{0.2}Co_{0.2}O_2$	ENF Technology Co., Ltd. (Asan, Korea)	50	25	15 days	3–4.4	1	100	43		[15]
$LiNi_{0.6}Mn_{0.2}Co_{0.2}O_2$	Beijing Easpring Material Technology Co., Ltd. (China)	80	55	7 days	3–4.4	0.1	100	54		[16]
$LiNi_{0.8}Mn_{0.1}Co_{0.1}O_2$	Umicore (Belgium)	20–25	30–50	3 months	3–4	1	300	97		[17]
$LiNi_{0.8}Co_{0.15}Al_{0.05}O_2$	Fuji CA1505	Open air	Ambient temperature	2 years	3–4.1	0.1	3	40		[18]
$LiNi_{0.8}Co_{0.15}Al_{0.05}O_2$	Hunan Changyuan Lico Co., Ltd	60	45	30 days	3–4.5	3	100	68.9		[19]

(continued)

Table 3.1 (continued)

Cathode materials	Synthesis condition of cathode materials	Humid atmosphere		Capacity fading					References
		relative humidity (%)	Temperature (°C)	Storage period	Cutoff voltage (V vs. Li/Li$^+$)	C-rate (C)	Cycle number	Capacity retention[1] (%)	
$LiNi_{0.8}Co_{0.2}O_2$	Synthesized by a sol–gel method using citric acid as a chelating agent	In air	Room temperature	6 months	3.0–4.2	0.1	1	79	[20]
$LiNi_{0.75}Ti_{0.05}Co_{0.2}O_2$	Synthesized by a sol–gel method using citric acid as a chelating agent	In air	Room temperature	6 months	3.0–4.2	0.1	1	89	[20]
$LiNiO_2$	Synthesized by a sol–gel method using citric acid as a chelating agent	In air	Room temperature	6 months	2.7–4.3	0.1	1	44	[21]

(continued)

Table 3.1 (continued)

Cathode materials	Synthesis condition of cathode materials	Humid atmosphere				Capacity fading				References
		relative humidity (%)	Temperature (°C)	Storage period	Cutoff voltage (V vs. Li/Li$^+$)	C-rate (C)	Cycle number	Capacity retention[1] (%)		
Carbon-coated LiFePO$_4$	Prepared from FePO$_4$(H$_2$O)$_2$ and Li$_2$CO$_3$. The stoichiometric amounts of precursors were thoroughly mixed in isopropanol. After drying, the blend was heated at 700 °C under a reducing atmosphere, i.e., argon + 5% hydrogen, for 6 h	55	21	6 months	2.2–4	0.1	1	56		[22]
LiFeSO$_4$F	FeSO$_4$·H$_2$O and LiF precursors were used to prepare tavorite LiFeSO$_4$F	85	20	2 h	2.3–4.5	0.1	2	80		[23]

(continued)

Table 3.1 (continued)

Cathode materials	Synthesis condition of cathode materials	Humid atmosphere			Capacity fading					References
		relative humidity (%)	Temperature (°C)	Storage period	Cutoff voltage (V vs. Li/Li$^+$)	C-rate (C)	Cycle number	Capacity retention[1] (%)		
Li$_2$MoO$_3$	Li$_2$MoO$_3$ powder was prepared by reducing commercial Li$_2$MoO$_4$ (Alfa Aesar) at 650 °C for 24 h in flowing H$_2$/Ar (10:90 v/v)	20	Room temperature	120 days	2.0–4.5	0.1	5	0		[24]

Fig. 3.2 Growth kinetics of interfacial reaction layers. Measured LiOH passivation layer thickness with atomic–resolution, time-resolved in situ environmental transmission electron microscopy as a function of exposure time for **a–c** NMC333, **d–f** NMC622 and **g–i** NMC811 at $P_{H2O} = 5 \times 10^{-2}$ Torr and room temperature. Reproduced with permission [25]. Copyright 2020, Nature Publishing Group

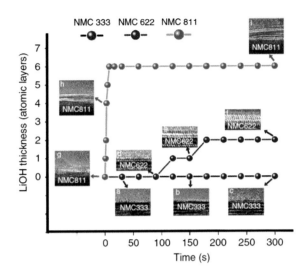

LiOH, lithium bicarbonate (LiHCO$_3$) and Li$_2$CO$_3$ adversely affect electrochemical performance.

Free energies for proton-for-lithium-ion-exchange reactions in cathode materials were calculated with first-principal calculations by Benedek et al. [27]. Protonation is most energetically favorable in layered systems, such as LiCoO$_2$ and Li$_2$MnO$_3$, and ion exchange in spinel LiMn$_2$O$_4$ and LiV$_3$O$_8$ are less favorable are.

The substitution of protons for Li$^+$ ions in LFP is unfavorable. The order of free energy for the protonation reaction is in agreement with the order of stability against moisture, as shown in Table 3.1 and in the results shown in Chap. 7. The relationship between the reaction free energy and the volume change in the reaction showed rough linearity (Fig. 3.3). The reason for the relationship is related to the flexibility to accommodate bonding of either lithium or proton layers, without severe distortion of the metal oxide octahedra, by adjustment of the interlayer spacings and/or by sliding of the layers relative to each other. In fact, under moisture, proton exchange, delithiation and redox reactions occur concertedly. On LiNiO$_2$-based cathode materials, the reduction of Ni^{3+} to Ni^{2+} and oxidation of lattice oxygen O^{2-} to active oxygen O$^-$ are accompanied by proton exchange. Continuously, oxidation of the reaction oxygen species on the surface of LiNiO$_2$ proceeds, finally forming molecular oxygen (O$_2$) (Eqs. 3.5–3.8) [21]. As a total reaction of LiNi$_{0.8}$Co$_{0.2}$O$_2$ in the case of contacting moisture, for example, Liu et al. proposed the following reaction shown in Eq. (3.9), which occurs on the surface of cathode materials [20].

$$Ni^{3+} + O^{2-} (\text{lattice}) \rightarrow Ni^{2+} + O^- (\text{active}) \tag{3.5}$$

$$O^- + O^- \rightarrow O^{2-} (\text{active}) + O \tag{3.6}$$

Fig. 3.3 Standard reaction free energy for the exchange of protons for Li versus volume change (difference between volume of protonated and lithiated systems) [27]. Reprinted with permission from Ref. [27]. Copyright 2008 American Chemical Society

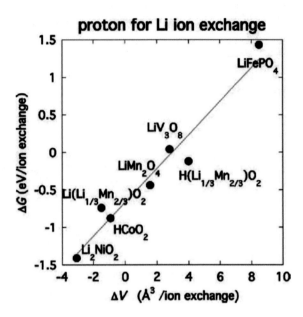

$$O^- + O \rightarrow O_2^- \text{ (active)} \tag{3.7}$$

$$O + O \rightarrow O_2 \uparrow \tag{3.8}$$

$$LiNi_{0.8}Co_{0.2}O_2 + 0.4H_2O \rightarrow 0.8LiOH + 0.8NiO + 0.2LiCoO_2 + 0.2O_2 \uparrow \tag{3.9}$$

In the oxidation state of Co ions in $LiNi_{0.8}Co_{0.2}O_2$ after contact with moisture, Co $2p$ X-ray photoelectron spectroscopy (XPS) spectra did not show any peak shift or decrease in intensity, indicating that Co ions are kept in trivalence before and after contact with moisture [20]. The authors considered that, based on the results, the poorer storage properties of $LiNiO_2$-based materials than those of $LiCoO_2$ most likely originate from the lower chemical stability of Ni^{3+} in the layered metal oxide structures.

Simulation results are available proposing the dissociative adsorption of H_2O to oxygen vacancies on the surface of $LiCoO_2$ as a trigger for the dissolution of transition metal ions from $LiCoO_2$ surfaces. The dissociative adsorption of H_2O is followed by electron transfer to Co ions on neighboring sites, producing Co^{2+} ions from Co^{3+} in $LiCoO_2$. The Co^{2+} ions are dissolved from the $LiCoO_2$ surface [28]. During acid treatment of Li_2MnO_3 (i.e., $Li[Li_{0.33}Mn_{0.67}]O_2$)-layered cathode materials, the phenomenon in which Li^+ ions are initially progressively removed from the lithium layers while Li^+ ions remain in the Li/Mn layers was observed in experiments with ^6Li and ^2H MAS NMR in conjunction with X-ray diffraction. In the case of prolonged acid leaching, a decrease in both the lithium and deuterium contents could be observed. As an explanation, lithia (Li_2O) dissolution from the

Fig. 3.4 Surface
disproportionation of Mn^{3+}
into Mn^{2+} and Mn^{4+}
followed by dissolution of
Mn^{2+} from the surface.
Reproduced with permission
from Ref. [31]

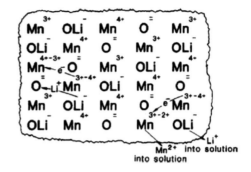

$Li[Li_{0.33}Mn_{0.67}]O_2$ surfaces in conjunction with ion exchange was considered [29]. Spinel lithium manganese oxide ($LiMn_2O_4$) is a promising cathode material that provides a high voltage of approximately 4 V and rate capability and is an abundant and nontoxic material. However, the $LiMn_2O_4$ cathode material undergoes capacity fading with charging/discharging cycles due to the dissolution of Mn ions from the $LiMn_2O_4$ surface [30]. Even upon contact of $LiMn_2O_4$ with moisture and aqueous electrolytes, dissolution of Mn ions occurs on the surface of $LiMn_2O_4$. Hunter's examinations provide pioneering information on Mn dissolution from $LiMn_2O_4$ (Fig. 3.4) [31].

The dissolution of Mn is considered to occur due to a disproportionation of Mn^{3+} into Mn^{4+} and Mn^{2+}. Under the operation of LIBs, in the presence of trace amounts of protons originating from ppm levels of H_2O, which generates HF by reacting with the $LiPF_6$ salt in the electrolyte, Eq. (3.10) occurs, resulting in the dissolution of Mn^{2+} into electrolytes or aqueous electrolytes [32].

$$2LiMn_2O_4(s) + 4H^+(aq.) \rightarrow 3MnO_2(s) + Mn^{2+}(aq.) + 2Li^+(aq.) + 2H_2O(aq.)$$
$$(3.10)$$

MnO_2 remains in the solid on the $LiMn_2O_4$ surfaces with maintenance of the spinel framework structure in which the $8a$ tetrahedral sites occupied by Li^+ ions in $LiMn_2O_4$ are completely empty [33]. Manthiram et al. investigated the amount of lithium extracted from $LiMn_{2-x}M_xO_4$ (LMMO, M = Cr, Fe, Co and Ni) spinel oxides to establish the lithium extraction mechanism [33]. Based on the results of X-ray diffraction and chemical composition analysis, it was found that lithium extraction from $LiMn_2O_4$ occurs by disproportionation of Mn^{3+} ions in $LiMn_2O_4$ into Mn^{4+} and Mn^{2+} ions, which was proposed by Hunter [31] and that the amount of lithium extraction is proportional to the Mn^{3+} content in LMMO. Because other M^{3+} ions in LMMO do not disproportionate into M^{4+} and M^{2+} ions, M ions in LMMO do not dissolve into electrolytes and aqueous electrolytes, and the amount of lithium extraction subsequently decreases. In addition, the discussion that ion exchange of Li^+ ions by H^+ ions does not occur in $LiMn_2O_4$ and LMMO because H^+ ions may not be stabilized in the tetrahedral site of the spinel structure was mentioned in their paper. Several examinations with simulation technologies of the aqueous dissolution of $LiMn_2O_4$ have been reported [34–36]. Benedek et al. calculated the free energies of Mn^{3+}, O^{2-} and Li^+ required to break the chemical bonds in the model of $LiMn_2O_4$, with orientations (001) and (110), embedded in a cell with 20 Å H_2O channels between periodically repeated slabs [34]. The results of the simulation showed that the bond-breaking energies of Li^+ ions were smallest, and those of Mn ions were largest. Li^+ ions may have a greater propensity for dissolution than other components in $LiMn_2O_4$. O^{2-} ion dissolution proceeded (fourfold coordination) with Mn ion dissolution. Conversely, the opposite was found to be true for (110) surfaces on which Mn was threefold coordinated. In Fig. 3.5, snapshots of the dissolution of Mn from the Mn−O terminated (001) slab are shown. In the time course shown in Fig. 3.5, bond breaking was sequential and ordered from weakest to strongest, and the breaking of the solute bonds with the substrate during dissolution was complemented by the increased hydration of the solute. In simulations under actual conditions in

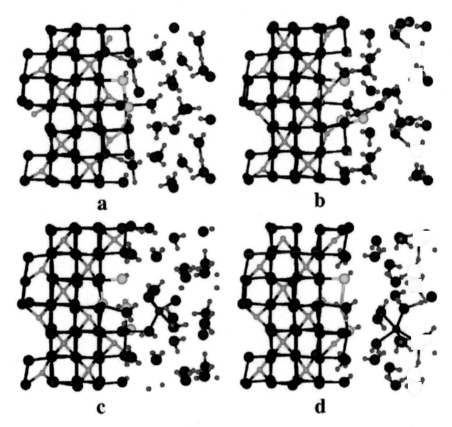

Fig. 3.5 Snapshots of the dissolution of Mn from the Mn−O terminated (001) slab. The dissolving Mn ion is dark blue, and coordinated O ions are yellow. Li^+ ions, Mn in bulk material and O ions and protons in light green, purple, red and light blue, respectively. The time course of simulation for Mn dissolution is a → b → c → d [34]. Reprinted with permission from Ref. [34]. Copyright 2012 American Chemical Society

LIBs, HF-promoted dissolution of Mn^{3+} ions from the $LiMn_2O_4$ surface [35] and the contribution of chemical reactions of adsorbed/decomposed organic fragments in SEI on anodes to dissolution [36] were also considered.

On the surfaces of cathode materials with high Ni content, LiOH and Li_2CO_3 are formed due to strong adsorption of H_2O, leaching of Li^+ ions (Li loss in cathode materials) from the surface of the cathode material and attack of CO_2 on the cathode surface, resulting in capacity loss of the cathode materials.

Washing is a commonly used method to remove the surface impurities of LiOH and Li_2CO_3 with H_2O [37, 38], alcohol [39] and polyaniline [40]. Kim et al. reported that although as-prepared $LiNi_{0.83}Co_{0.15}Al_{0.02}O_2$ exhibited large capacity fading with charge/discharge cycles, two washes with H_2O and subsequent heat treatment of the water-treated $LiNi_{0.83}Co_{0.15}Al_{0.02}O_2$ improved the cyclability of the treated sample [37]. Lee et al. examined the crystal and electronic structures of $LiNi_{0.88}Co_{0.11}Al_{0.01}O_2$ after washing with H_2O and heat treatment, and they reported that the treated $LiNi_{0.88}Co_{0.11}Al_{0.01}O_2$ had a favorable environment for Li^+ ions to diffuse, a lower increment in cation disorder and a better local structural retentivity after 300 cycles [38]. However, the result showed that the treated $LiNi_{0.8}Co_{0.1}Mn_{0.1}O_2$ sample, which could exhibit the improved cycle performance after washing and heat treatment, was more vulnerable to moisture than the as-prepared sample [41]. Pritzl et al. reported that in $LiNi_{0.85}Co_{0.10}Mn_{0.05}O_2$, the positive effect of washing/drying on capacity retention could not be observed because nickel oxides such as NiOOH and NiO formed continuously during washing due to Li^+/H^+ ion exchange near the surface of $LiNi_{0.85}Co_{0.10}Mn_{0.05}O_2$ to block Li^+ migration to $LiNi_{0.85}Co_{0.10}Mn_{0.05}O_2$, which suggested that the drying temperature was a critical step [42]. Jeong et al. also reported the formation of cobalt oxide on the surface of $LiCoO_2$ after washing it with H_2O, and in their case, the thickness of the cobalt oxide layer and the type of the materials were changed by the temperature of the heat treatment after the washing process [43]. The volume of H_2O, period of washing, temperature of heat treatment and period of heat treatment should be noted to form resistance components on the surface of samples by washing with H_2O to remove LiOH and Li_2CO_3.

To protect the surface of the cathode materials against moisture, a surface coating was applied to the surface of the cathode materials. As examples of coating layers with additional materials, Mn-based spinel oxide $LiMn_{1.9}Al_{0.1}O_4$ layers [44] and hydrophobic self-assembled monolayers of octadecyl phosphate [45] formed on Ni-rich layered oxides and chemically grafted trifluoromethylphenyl groups formed on LFP [46] are well known. In addition, on Ni-rich layered oxides, substituting a small amount of Al for Ni in the crystal lattice improves the chemical stability against moisture by inhibiting the formation of LiOH, Li_2CO_3, $LiHCO_3$ and NiO, which is caused by contact with H_2O, in the near-surface region [47]. Moreover, as mentioned above, washing and heat treatment of the cathode material of $LiCoO_2$ produces coating layers of CoO(OH), CoO and Co_3O_4, which are self-generated and waterproof properties [43]. These coating technologies were developed and applied to H_2O processes for water-based binders to produce electrodes in LIBs, as will be mentioned below.

3.3 Sensitivity of Separators to Humidity and Protection Against Humidity

Separators are located between cathodes and anodes to prevent electrical short circuits between them. The separators also serve as electrolyte reservoirs for the transport of Li^+ ions and should not block the transport of Li^+ ions in the separators to maintain the high performance of cathodes and anodes. In addition, separators should have the property of being able to shut the battery down when overheating occurs in the cell to maintain battery safety. In the present batteries, microporous polymer membranes, multilayer and modified membranes, nonwoven mats and composite membranes combined with inorganic particles are developed and used [48]. H_2O adsorbed on the separators contaminates LIB cells and degrades battery performance. The polypropylene separator contains a very small amount of H_2O, while the glass fiber separator contains a very large amount of H_2O and requires a long drying time to remove H_2O from the separator [49]. Recently, separators have been shown to have moisture repulsion property as well as mechanical and Li^+ ion transportation properties and shutdown functions.

Polymer membranes are combined with inorganic particles such as alumina (Al_2O_3) and silica (SiO_2) to acquire mechanical and thermal properties. Although the coating of Al_2O_3 and SiO_2 on separators utilizes coating solutions that contain polymeric binders dissolved in an organic solvent, to reduce the production cost in the production processes of composite membrane separators, recently, water processes have been applied for composite membrane separators.

H_2O adsorption on the composite membrane separators should be reduced based on the design of the separator surface. Cho et al. proposed a water-repulsion surface of membrane separators inspired by a plant leaf surface [50]. Emulsion wax domains were dispersed with Al_2O_3 and binders on the membrane separators (Fig. 3.6). Yang et al. also proposed the structure of silicone nanofilaments (SNFs)/polypropylene separator (Celgard 2400)/SNFs for superliquid electrolyte (LE)-philic/superhydrophobic separators. The separator showed features of low moisture uptake (0%), fast liquid electrolyte (LE) diffusion (454 ms), high LE uptake (287.8%), LE retention rate and Li^+ conductivity [51].

Separators that function as scavengers for H_2O were also developed. Molecular sieves that capture H_2O contamination by 4-Å sieves were modified on a functionalized poly(vinylidene fluoride-cohexafluoropropylene)@polyacrylonitrile (PVdF-HFP@PAN) separator [52]. The capture of HF produced by the reaction of $LiPF_6$ and H_2O in electrolytes was also designed for functional separators. The separator membranes consisted of a 25% cross-linked divinylbenzene backbone functionalized with 4-vinylpyridine [53].

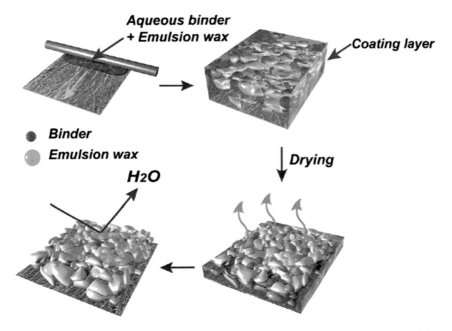

Fig. 3.6 Schematic of the fabrication process for CCS from aqueous coating solution with emulsion wax [50]. Reproduced from Ref. [50] with permission from the Royal Society of Chemistry

3.4 Sensitivity of Electrolytes to Humidity and Protection Against Humidity

It is well known that the H_2O content in electrolytes affects the Coulombic efficiency of Li deposition/dissolution during charging/discharging cycles at Li metal anodes. Osaka et al. examined the effect of H_2O in electrolytes such as 1 M $LiPF_6$/EC + DEC [54] and 1 M lithium bis(trifluoromethanesulfonyl)imide (LiTFSI)/dimethyl sulfoxide (DMSO) [55] on the Coulombic efficiency of the deposition/dissolution of Li. In the presence of CO_2 in the electrolyte of 1 M $LiPF_6$/EC + DEC, the Coulombic efficiency of deposition/dissolution of Li increased with an increase in electrolyte H_2O content of up to 35 ppm and reached a maximum of 88.9% (Fig. 3.7).

After reaching the maximum, the Coulombic efficiency decreased gradually with an increase in H_2O content [54]. In addition, surface analysis of Li deposited after the 1st deposition process with XPS revealed the composition of SEI formed on the Li surface. At low H_2O contents, lithium alkyl carbonate ($ROCO_2Li$) and Li_2CO_3 were formed as the main components in the outer and inner layers of the SEI, respectively. Conversely, around the H_2O content of 35 ppm, Li_2CO_3 was a surface component of SEI, and lithium oxide (Li_2O) was formed in the SEI layer. Li_2CO_3 was formed in the reaction of $ROCO_2Li$ with H_2O (Eq. 3.11). Li_2O was formed by the reaction between H_2O and Li^+ (Eq. 3.12).

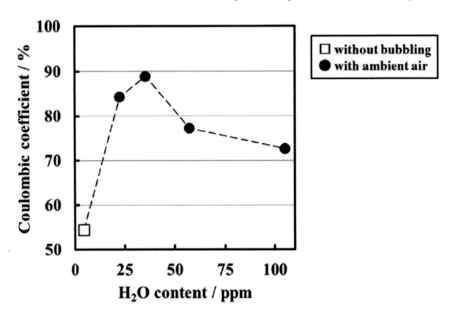

Fig. 3.7 Effects of H_2O content in the electrolyte of 1 M $LiPF_6$/EC + DEC bubbled on the coulombic efficiency of a lithium metal anode [54]. Reprinted from J. Power Sources, 261, Togasaki, N.; Momma, T.; Osaka, T.: Enhancement effect of trace H_2O on the charge–discharge cycling performance of a Li metal anode, 23–27, Copyright (2014), with permission from Elsevier

$$2ROCO_2Li + H_2O \rightarrow Li_2CO_3 + 2ROH + CO_2 \qquad (3.11)$$

$$LiOH + Li^+ + e^- \rightarrow Li_2O + H_2 \qquad (3.12)$$

Li_2CO_3 can protect lithium metal from side reactions with the electrolyte during charging–discharging cycles and, therefore, retain the high Coulombic efficiency. At H_2O contents over 35 ppm, thick lithium fluoride (LiF) layers were formed on the surface of the SEI.

$$PF_6^- + 3Li^+ + 2e^- \rightarrow 3LiF \downarrow + PF_3 \qquad (3.13)$$

$$ROCO_2Li + HF \rightarrow LiF \downarrow + ROH + CO_2 \qquad (3.14)$$

$$Li_2CO_3 + 2HF \rightarrow 2LiF \downarrow + H_2CO_3 \qquad (3.15)$$

$$Li_2O + 2HF \rightarrow 2LiF \downarrow + H_2O \qquad (3.16)$$

HF in Eq. (3.14–3.16) was formed in Eqs. (1.1 and 1.2) of the reaction between H_2O and $LiPF_6$. LiF can also protect lithium metal from side reactions with the electrolyte during charging–discharging cycles. However, the ionic conductivity of LiF is much smaller than that of Li_2CO_3. Therefore, in the deposition/dissolution of Li on the LiF layer, a higher overpotential was observed, resulting in a decrease in the Coulombic efficiency [54].

The dependence of the Coulombic efficiency on H_2O content in the electrolytes of 1 M LiTFSI/DMSO exhibited different behaviors [55]. Up to a H_2O content of 1000 ppm in the presence of CO_2, a high Coulombic efficiency of over 80% was retained because Li_2CO_3 and Li_2O were formed on the SEI surfaces. The optimal H_2O content for high Coulombic efficiency of the deposition/dissolution of Li was examined with an electrolyte of 1 M LiTFSI/tetraethylene glycol dimethyl ether (TEGDME). In this research, the presence/absence of CO_2 in the electrolyte is not discussed. The maximum Coulombic efficiency of approximately 80% was observed at a H_2O content of 1000 ppm. The high Coulombic efficiency was associated with the increase in the composition ratio of Li_2O in the SEI layer. At H_2O contents over 1000 ppm, the formation of resistive LiOH induced the local deposition of Li and resulted in low Coulombic efficiency [56]. LiOH was formed upon reduction of H_2O on anodes (Eqs. 3.17 and 3.18), which precipitated on the anode surface and acted as a blocking agent with high interfacial resistance [57].

$$H_2O + e^- \rightarrow OH^- + 1/2 \; H_2 \uparrow \qquad (3.17)$$

$$Li^+ + OH^- \rightarrow LiOH \downarrow \qquad (3.18)$$

EC, which is one of the components of electrolytes for LIBs, is chemically decomposed by H_2O or hydroxide (OH^-), which is produced in Eq. (3.17) to form CO_2, ethylene glycol and polyethylene glycol. In 1.5 M $LiClO_4$/EC electrolyte at 40 °C the decomposition rates of EC at H_2O contents of <20, 1000 and 5000 ppm were approximately 10^{-12}, 10^{-11} and 10^{-10} $mol_{CO2}/(s \; g(EC))$, respectively. At 5000 ppm H_2O and 167 ppm OH^-, the rate became 10^{-9} $mol_{CO2}/(s \; g(EC))$. The rate constants were evaluated based on the rate of CO_2 production from EC. EC decomposition was enhanced by the presence of OH^- in the electrolytes [58]. The H_2 gas produced in Eq. (3.17) is one of the gases that swells pouch bag cells of LIBs after long-term cycling or storage at an elevated temperature [59]. SEI formation on graphite anode surfaces by reduction of electrolyte components lowers the gas formation of H_2; in contrast, LTO anode surfaces, on which an SEI is not formed because the anode potential is too high to reduce electrolyte components, produce excessive H_2 gas [60].

Lithium salts such as $LiPF_6$, LiTFSI, lithium percholate ($LiClO_4$), lithium tetrafluoroborporate ($LiBF_4$), lithium bis(fluorosulfonyl)imide (LiFSI), lithium hexafluoroarsenate(V) ($LiAsF_6$) and lithium bis(oxalato)borate (LiBOB) are used in LIBs. A close look at lithium-conducting salts clearly reveals the difference in the stability of the lithium salts against moisture. As mentioned above, $LiPF_6$ reacts with H_2O to form HF, and several serious problems subsequently occur in LIBs. The rate of the reaction between $LiPF_6$ and H_2O was measured in 1 M $LiPF_6$ in a mixture of EC/EMC (3:7, v/v) after the addition of 0.3% H_2O. Almost all the H_2O in the electrolyte was consumed in the reaction with $LiPF_6$ in 5 days [61]. The reaction between $LiPF_6$ and H_2O started ring-opening polymerization of EC and CO_2 gas [58]. Relative rates of EC loss with various salts in a mixture of EC/EMC (1:1 mol ratio) were compared. The relative rates were in the order of $LiPF_6 > LiBF_4 > LiAsF_6 \gg LiTFSI$. The electrolyte containing LiTFSI was stable, and EC loss could not be observed [62]. A cell prepared with a $LiBF_4$ salt that was less moisture sensitive, thermally stable and could form improved passivation of the Al current collector [63] could perform even under conditions of a H_2O content of 620 ppm at room temperature. At 60 °C a cell with the $LiBF_4$ electrolyte containing 80 ppm H_2O did not suffer from fading battery performance [64]. Cui et al. proposed preparation and H_2O removal methods for $LiBF_4$ [65]. The H_2O content in $LiBF_4$ dried at 155 °C decreased to 55.6 ppm. Drying $LiBF_4$ above 155 °C produced the decomposition products of $LiBF_4$: LiF and BF_2OH. A Li/mesophase carbon microbead (MCMB) half-cell with $LiBF_4$ as the electrolyte salt showed discharge plateaus at approximately 1.5 V versus Li/Li^+, which originated from the formation of SEI on the MCMB surface and deteriorated the cycling performance due to the thicker and more resistant SEI even at an H_2O content of 55.6 ppm.

Bund et al. examined the mechanism of hydrolysis of $LiPF_6$ in EC/DEC (1:1, v/v) on a time scale of several weeks with ion chromatography, coulometric Karl Fischer titration and acid-base titration. Based on the stoichiometry of the decomposition reaction of $LiPF_6$, they added the reaction shown in Eq. (3.19), which follows reactions in Eqs. (1.1) and (1.2) as the decomposition sequence of $LiPF_6$.

$$POF_3 + H_2O \rightarrow HF + HPO_2F_2 \qquad (3.19)$$

The concentration ratio between HF and HPO_2F_2 throughout the experiment was 3:1, and only minimal amounts of POF_3 present in the electrolyte strongly indicated that reaction (21) proceeded.

$$HPO_2F_2 + H_2O \rightarrow H_2PO_3F + HF \qquad (3.20)$$

$$H_2PO_3F + H_2O \rightarrow H_3PO_4 + HF \qquad (3.21)$$

Although reactions (3.20) and (3.21) may be considered to follow reaction (3.19), low concentrations of H_2O (1000 ppm and below H_2O content) did not lead to significant formation of H_2PO_3F in their experiments. With the kinetic equations for reactions (1.1), (1.2) and (3.19), fitting between experimental concentration versus time plots and the theoretical concentration *vs.* time plot succeeded well. The rate constants for each reaction could be calculated [66]. Nowak et al. categorized the influence of HF formed in the reaction of $LiPF_6$ and H_2O to cell components (Fig. 3.8) [67]. Graphite, CB, and poly(vinylidine difluoride) (PVdF), among others, were not affected by HF. Oxidized Si nanopowder and LiOH-containing LFP produced H_2O in the reaction with HF, and then, the H_2O produced attacks on $LiPF_6$ resulted

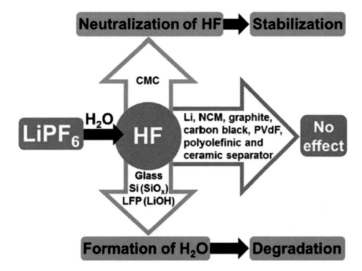

Fig. 3.8 Schematic depicting different kinds of influences of cell components on the thermal stability of $LiPF_6$ in organic carbonates dependent on the reactivity against HF [67]. Reprinted from Electrochim. Acta, 222, Wiemers-Meyer, S.; Jeremias, S.; Winter, M.; Nowak, S.: Influence of battery cell components and water on the thermal and chemical stability of $LiPF_6$-based lithium-ion battery electrolytes, 1267–1271, Copyright (2016), with permission from Elsevier

in degradation of the Li^+ conductivity of the electrolytes. Sodium carboxymethyl cellulose (CMC-Na) could neutralize HF and stabilize HF.

LiBOB has received significant attention, as LiBOB has better water tolerance than $LiPF_6$-based systems. Feng et al. examined the battery performance of $LiMn_2O_4$/Li cells using $LiPF_6$-EC/DEC and LiBOB-tetramethylene sulfone (SL)/DEC electrolytes with controlled H_2O concentrations [68]. In their results, when the H_2O content was less than 300 ppm, the cycle and rate performances of the cells containing the $LiPF_6$-based electrolyte were better than those of the cells of the LiBOB-based electrolyte because $LiPF_6$ does not react with H_2O and $LiPF_6$ can exhibit inherent properties as a lithium salt. However, at H_2O contents over 300 ppm, the cells of the LiBOB-based electrolyte were better than those of the $LiPF_6$-based electrolyte. In addition, the dissolution of Mn ions from $LiMn_2O_4$ cathode materials was confirmed to be sufficiently prevented in LiBOB-based electrolytes when compared with $LiPF_6$-based electrolytes [68]. The water tolerance of LiBOB-based electrolytes was based on the hydrolytic reaction of LiBOB with H_2O [69]. Although LiBOB electrolytes containing trace amounts (~100 ppm) of moisture appear relatively stable, higher moisture contents (~1 wt.%) lead to observable salt decomposition, resulting in the generation of $B(C_2O_4)(OH)$ and $LiB(C_2O_4)(OH)_2$ compounds that do not dissolve in electrolytes and degrade battery performance [70].

References

1. Stich M, Pandey N, Bund A (2017) J Power Sources 364:84
2. Eser JC, Wirsching T, Weidler PG, Altvater A, Börnhorst T, Kumberg J, Schöne G, Müller M, Scharfer P, Schabel W (2020) Energy Technol 8:1801162
3. Wu Y, Jiang C, Wan C, Tsuchida E (2000) Electrochem Commun 2:626
4. Wu YP, Jiang C, Wan C, Holze R (2002) J Power Sources 112:255
5. Fu LJ, Zhang HP, Wu YP, Wu HQ, Holze R (2005) Electrochem Solid-State Lett 8:A456
6. Zhang Y, Lv W, Huang Z, Zhou G, Deng Y, Zhang J, Zhang C, Hao B, Qi Q, He YB, Kang F, Yang QH (2019) Sci Bull 64:910
7. Shen X, Li Y, Qian T, Liu J, Zhou J, Yan C, Goodenough JB (2009) Nat Commun 10:900
8. Xiao Y, Xu R, Yan C, Liang Y, Ding JF, Huang JQ (2020) Sci Bull 65:909
9. Zhao J, Zhou G, Yan K, Xie J, Li Y, Liao L, Jin Y, Liu K, Hsu PC, Wang J, Cheng HM, Cui Y (2017) Nat Nanotech 12:993
10. Liao K, Wu S, Mu X, Lu Q, Han M, He P, Shao Z, Zhou H (2018) Adv Mater 30:1705711
11. Qiana J, Xu W, Bhattacharya P, Engelhard M, Henderson WA, Zhang Y, Zhang JG (2015) Nano Energy 15:135
12. Mehdi BL, Stevens A, Qian J, Park C, Xu W, Henderson WA, Zhang JG, Mueller KT, Browning ND (2016) Sci Rep 6:34267
13. Zhang X, Jiang WJ, Zhu XP, Mauger AQ, Julien CM (2011) J Power Sources 196:5102
14. Huang B, Qian K, Liu Y, Liu D, Zhou K, Kang F, Li B (2019) ACS Sustainable Chem Eng 7:7378
15. Park JH, Park JK, Lee JW (2016) Bull Korean Chem Soc 37:344
16. Chen Z, Wang J, Huang J, Fu T, Sun G, Lai S, Zhou R, Li K, Zhao J (2017) J Power Sources 363:168
17. Jung R, Morasch R, Karayaylali P, Phillips K, Maglia F, Stinner C, Shao-Horn Y, Gasteiger HA (2018) J Electrochem Soc 165(2):A132
18. Zhuang GV, Chen G, Shim J, Song X, Ross PN, Richardson TJ (2004) J Power Sources 134:293
19. Wang C, Shao L, Guo X, Xi X, Yang L, Huang C, Zhou C, Zhao H, Yin D, Wang Z (2019) ACS Appl Mater Interfaces 11:44036
20. Liu H, Yang Y, Zhang J (2006) J Power Sources 162:644
21. Liu HS, Zhang ZR, Gong ZL, Yang Y (2004) Electrochem Solid-State Lett 7:A190
22. Zaghib K, Dontigny M, Charest P, Labrecque JF, Guerfi A, Kopec M, Mauger A, Gendron F, Julien CM (2008) J Power Sources 185:698
23. Zhang L, Tarascon JM, Sougrati MT, Rousse G, Chen G (2015) J Mater Chem A 3:16988
24. Ma J, Gao Y, Wang Z, Chen L (2014) J Power Sources 258:314
25. Zou L, He Y, Liu Z, Jia H, Zhu J, Zheng J, Wang G, Li X, Xiao J, Liu J, Zhang JG, Chen G, Wang C (2020) Nat Commun 11:3204
26. Shkrob IA, Gilbert JA, Phillips PJ, Klie R, Haasch RT, Bareño J, Abraham DP (2017) J Electrochem Soc 164:A1489
27. Benedek R, Thackeray MM, van de Walle (2008) Chem Mater 20: 5485
28. Cherkashinin G, Jaegermann W (2016) J Chem Phys 144: 184706
29. Paik Y, Grey CP, Johnson CS, Kim JS, Thackeray MM (2002) Chem Mater 14:5109
30. Xia YY, Zhou YH, Yoshio M (1997) J Electrochem Soc 144:2593
31. Hunter JC (1981) J Solid State Chem 39:142
32. Bhandari A, Bhattacharya J (2017) J Electrochem Soc 164:A106
33. Knight JC (2015) J Electrochem Soc 162:A426
34. Benedek R, Thackeray MM (2012) J Phys Chem C 116:4050
35. Benedek R (2017) J Phys Chem C 121:22049
36. Leung K (2017) Chem Mater 29:2550
37. Kim J, Hong Y, Ryu KS, Kim MG, Cho J (2006) Electrochem Solid-State Lett 9:A19
38. Lee W, Lee D, Kim Y, Choi W, Yoon WS (2020) J Mater Chem A 8:10206
39. Zheng X, Li X, Wang Z, Guo H, Huang Z, Yan G, Wang D (2016) Electrochim Acta 191:832

40. Xu S, Du C, Xu X, Han G, Zuo P, Cheng X, Ma Y, Yin G (2017) Electrochim Acta 248:534
41. Xiong X, Wang Z, Yue P, Guo H, Wu F, Wang J, Li X (2013) J Power Sources 222:318
42. Pritzl D, Teufl T, Freiberg ATS (2019) J Electrochem Soc 166:A4056
43. Jeong S, Kim J, Mun J (2019) J Electrochem Soc 166:A5038
44. Oh P, Song B, Li W, Manthiram A (2016) J Mater Chem A 4:5839
45. Gu W, Dong Q, Zheng L, Liu Y, Mao Y, Zhao Y, Duan W, Lin H, Shen Y, Chen L (2020) ACS Appl Mater Interfaces 12:1937
46. Delaporte N, Trudeau ML, Bélanger D, Zaghib K (2020) Materials 13:942
47. You Y, Celio H, Li J, Dolocan A, Manthiram A (2018) Angew Chem Int Ed 57:6480
48. Lee H, Yanilmaz M, Toprakci O, Fu K, Zhang X (2014) Energy Environ Sci 7:3857
49. Smith RS (1996) Advanced dry room concepts. In: First battery technology symposium on advanced secondary batteries, Seoul, Korea
50. Kim SW, Cho KY (2016) J Mater Chem A 4:5069
51. Yang Y, Li B, Li L, Seeger S, Zhang J (2019) iScience 16:420
52. Kang Y, Deng C, Wang Z, Chen Y, Liu X, Liang Z, Li T, Hu Q, Zhao Y (2020) Nanoscale Res Lett 15:107
53. Banerjee A, Ziv B, Shilina Y, Luski S, Aurbach D, Halalay IC (2017) ACS Energy Lett 2:2388
54. Togasaki N, Momma T, Osaka T (2014) J Power Sources 261:23
55. Togasaki N, Momma T, Osaka T (2015) J Power Sources 294:588
56. Koshikawa H, Matsuda S, Kamiya K, Kubo Y, Uosaki K, Hashimoto K, Nakanishi S (2017) J Power Sources 350:73
57. Arora P, White RE, Doyle M (1998) J Electrochem Soc 145:3647
58. Metzger M, Strehle B, Solchenbach S, Gasteiger HA (2016) J Electrochem Soc 163:A1219
59. Metzger M, Strehle B, Solchenbach S, Gasteiger HA (2016) J Electrochem Soc 163:A798
60. Bernhard R, Metzger M, Gasteiger HA (2015) J Electrochem Soc 162:A1984
61. Han HB, Zhou SS, Zhang DJ, Feng SW, Li LF, Liu K, Feng WF, Nie J, Li H, Huang XJ, Armand M, Zhou ZB (2011) J Power Sources 196:3623
62. Sloop SE, Kerr JB, Kinoshita K (2003) J Power Sources 119–121:330
63. Ellis LD, Hill IG, Gering KL, Dahn JR (2017) J Electrochem Soc 164(12):A2426
64. Zhang SS, Xu K, Jow TR (2002) J Electrochem Soc 149(5):A586
65. Zhao D, Lei D, Wang P, Li S, Zhang H, Cui X (2019) ChemistrySelect 4:5853
66. Stich M, Göttlinger M, Kurniawan M, Schmidt U, Bund A (2018) J Phys Chem C 122: 8836
67. Wiemers-Meyer S, Jeremias S, Winter M, Nowak S (2016) Electrochim Acta 222:1267
68. Cui X, Tang F, Zhang Y, Li C, Zhao D, Zhou F, Li S, Feng H (2018) Electrochim Acta 273:191
69. Li C, Li Z, Wang P, Liu H, Zhao D, Wang SX, Li S (2019) New J Chem 43:14238
70. Yang L, Furczon MM, Xiao A, Lucht BL, Zhang Z, Abraham DP (2010) J Power Sources 195:1698

Chapter 4
SEI and Water

Abstract The formation of solid electrolyte interphases (SEIs) is critical to ensure a safe operation and a long life of LIBs. H_2O plays the most important role in the formation of SEIs. H_2O initiates the formation of SEIs and is formed during the processes. In the viewpoint of the SEI formation, H_2O is not a nuisance and is essential to LIBs. In this chapter, the reaction mechanism of the SEI formation is summarized. The importance of H_2O is discussed.

Keywords Solid electrolyte interphase · H_2 evaluation · Self-catalyzed process · Water-driven hydrolysis · Activation energy

4.1 Importance of SEIs in LIBs

The passivation of anodes by electrolyte reduction products forming SEIs is critical to ensure a safe operation and a long life of the battery because the SEI halts further decomposition of the organic electrolyte [1–5]. Several SEI formation processes compete during the initial charging process of anodes. A large number of chemical compounds, such as POF_3, POF_2OH, Li_2O, LiF, LiOH, Li_2CO_3, alkoxides, alkyl carbonates and olefins, are formed on the surface of carbon anode materials. The formation of compounds starts by initiating reactions of lithium salts and solvents with H_2O. The POF_3, POF_2OH, Li_2O, LiF, LiOH and Li_2CO_3 compounds are formed in the reactions shown in Eqs. (1.2), (3.2), (3.4), (3.11), (3.12), (3.14), (3.15), (3.16), (3.18) and (3.19) [6]. The formation of alkoxides, alkyl carbonates and olefins, among others, is initiated by the reaction of OH^- with solvent, as shown in Fig. 4.1 [7]. OH^- are formed in the reduction of H_2O (Eq. 3.17). Additionally, on Li metal [8, 9] and silicon [10] anodes, the same chemical compounds are produced in the SEI according to the reactions mentioned above, although the ratios of components in the formed SEI are different.

F. Matsumoto and T. Gunji, *Water in Lithium-Ion Batteries*, SpringerBriefs in Energy,
https://doi.org/10.1007/978-981-16-8786-0_4

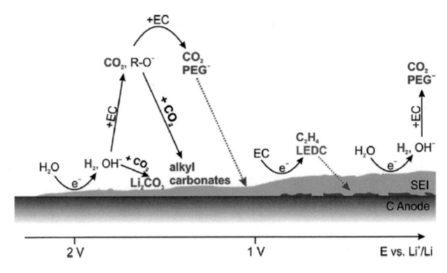

Fig. 4.1 Electrolyte decomposition and SEI formation on a carbon-based anode during the first polarization in a standard electrolyte with H_2O contamination. Reprinted with permission from Ref. [7]. Copyright 2020 American Chemical Society

4.2   Relationship Between the Elemental Content of the SEI and H_2O Content

The elemental content of the SEI might strongly depend on the H_2O concentration in the electrolyte. However, studies on the dependence of the elemental content of the SEI on the H_2O concentration in electrolytes could not be found, although the dependence of the elemental content of the SEI on the composition of electrolytes has been researched by many researchers, and many review papers have been published [3–5, 11]. Systematic research on the relationship between the SEI composition and H_2O content should be conducted. Only a few studies have examined the SEI and H_2O [12–14]. For example, Li et al. examined the influence of H_2O on the SEI formed in 0.7 M LiBOB − EC/DEC (1:1, by volume) electrolyte [14]. They concluded that the constitution of SEI depended on the H_2O content in the electrolyte, as shown in Fig. 4.2. Under conditions of a very low content of H_2O, inorganic components of Li_2CO_3 and Li_2O are formed on and near the surface of MCMB according to Eqs. (3.3), (3.4), (3.11), (3.12), (3.17), (3.18), (4.1) and (4.2).

$$Li_2O + CO_2 \rightarrow Li_2CO_3 \tag{4.1}$$

$$H_2O + EC + 2e^- + 2Li^+ \rightarrow LiOH + CH_3CH_2OLi + CO_2 \tag{4.2}$$

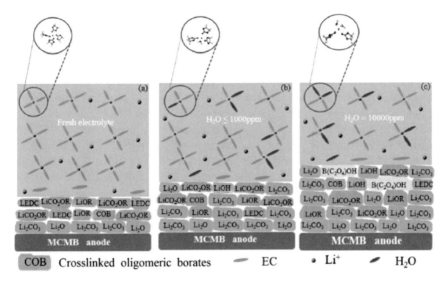

Fig. 4.2 Diagrammatic drawing of the destruction mechanism of H_2O in different electrolyte systems. Reprinted with permission from Ref. [14]. Copyright 2021 American Chemical Society

Solvated structures of $Li^+(EC)_2(H_2O)_2$ and $Li^+(EC)_3(H_2O)$ are sources of lithium alkoxide (4.2), lithium alkyl carbonate [15], lithium alkyl dicarbonate [15] and cross-linked oligometric borate (COB) components formed in the outer layer. When the H_2O content is less than 1000 ppm, excessive inorganic components that worsen battery performance are contained in the SEI. When the content of the additional H_2O reaches 10,000 ppm, insoluble $LiB(C_2O_4)(OH)_2$ and $B(C_2O_4)OH$ products are generated because of the hydrolytic reaction of LiBOB.

The relationship between the amount of gas evaluated and the H_2O content has been discussed because gas evolution during charging and discharging processes and storage in charged states causes degradation of battery performance due to cell swelling [16]. For example, in the case of EC-containing electrolytes, ethylene (C_2H_4) from the reduction of EC and hydrogen (H_2) generated in the reduction of H_2O (Eq. 3.17) and CO_2 formed in the EC hydrolysis reaction catalyzed by OH^- formed in Eq. (3.17) were evaluated. Joho et al. performed differential electrochemical mass spectrometry (DEMS) in a gas evaluation on graphite half-cells with different H_2O contents of 250, 1000 and 4000 ppm, and they found that the amount of C_2H_4 decreased with increasing H_2O and that the formation of H_2 increased with H_2O content [16]. Bernhard et al. also examined gas evaluation on a graphite surface on which SEI was and was not formed in a 2% vinylene carbonate (VC)-containing electrolyte [17]. On the surface of the graphite in electrolyte containing 4000 ppm H_2O, during the first charging process, H_2 evaluation occurred at an evaluation rate of 4 ppm s^{-1}. H_2 evaluation occurred in the following charging process, although the amount of H_2 evaluation gradually decreased. Conversely, on the SEI-formed graphite that was formed in an electrolyte containing VC (<20 ppm H_2O), H_2 and

C_2H_4 evolution during the charging process could be reduced even in electrolytes containing 4000 ppm H_2O. This result supports the importance of SEI formation with VC on graphite to effectively lower the gassing rate of the electrode during cycling.

The SEI formed on graphite electrodes prepared with H_2O process using water-based binders, which will be mentioned below, was largely comparable to that formed on graphite electrode prepared with a conventional PVdF binder. Jeschull et al. examined the SEI composition formed on graphite anodes prepared with water-based binders: CMC-Na, SBR and poly(sodium acrylate) (PAA-Na) [18]. The thickness of the SEI layers estimated with XPS was smaller than 11 nm for all graphite electrodes prepared with water-based binders of PAA-Na, CMC-Na:SBR and CMC-Na:PAA-Na, and the SEI layer thickness for the graphite:PVdF electrode was estimated to be 15 nm. The reason for the thinner SEI layer observed with PAA-Na, CMC-Na:SBR and CMC-Na:PAA-Na binders was that notable amounts of carboxylates and alkoxides on the surfaces of graphite electrodes prepared with water-based binders were formed when compared with those prepared with PVdF binder. In addition, carboxylate- and alkoxide-rich SEI layers could protect the graphite surface from irreversible reactions, resulting in higher anode performance of water-based binder/graphite anodes than a PVdF/graphite anode [19].

Cathode/electrolyte interphases (CEIs) are formed on the cathode surfaces during charging/discharging processes. The CEIs originate from the reaction of the active cathode surface with electrolyte solvents, conducting salt, hydrolysis products such as HF and the reactions of additives to form a stable layer that protects cathode surfaces as the SEI is formed [20–23]. For an example, POF_3 formed in Eqs. (1.1) and (1.2) reacts with dialkyl carbonates to produce dialkyl fluorophosphates, and the cathode covered with the products degrades the cycle performance [24]. The reaction of an electrolyte of 1 M $LiPF_6$ in EC/DMC/DEC (1:1:1) on a $LiNi_{0.5}Mn_{1.5}O_2$ cathode surface in the charging process at 4–5.3 V (vs. Li) produced poly(ethylenecarbonate) on the surface [25]. HF formed upon H_2O contamination of the electrolyte reacted with the V_2O_5 cathode through a self-catalyzed process. The products formed through the self-catalyzed process caused active mass loss and large capacity degradation during discharging/charging processes. In the catalytic reaction, H_2O was consumed and formed (Fig. 4.3) [26].

4.3   Gas Formation Related to H_2O

As a side reaction on cathodes, the release of reactive oxygen at higher states of charge of cathodes was observed and the electrolytes reacted with reactive oxygen. Among the reactions, H_2O was produced, which might become a source of the decomposition reaction of $LiPF_6$. Gray et al. carefully summarized the electrolyte oxidation pathways on cathode surfaces (Fig. 4.4) [27]. Metzger et al. examined anodic oxidation of conductive carbon in cathodes and EC at high voltage with online electrochemical mass spectrometry [28]. They evaluated water-driven hydrolysis of

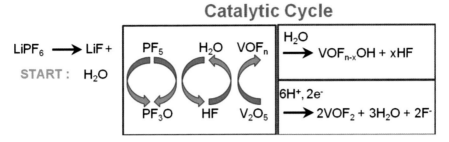

Fig. 4.3 Simplified mechanism for the reaction of $LiPF_6$ with H_2O and the catalytic reaction cycle of HF in the presence of V_2O_5. Reprinted with permission from Ref. [26]. Copyright 2012 American Chemical Society

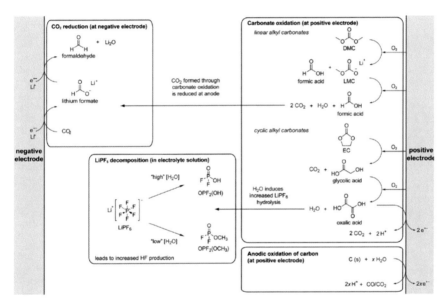

Fig. 4.4 Overview of electrolyte decomposition reactions that occur at high voltages (or high SOC) initiated at the positive electrode. Reprinted with permission from Ref. [27]. Copyright 2020 American Chemical Society

EC (Eq. 4.3) based on analysis of the amount of CO_2 formed during the chemical oxidation of EC in 1.5 M $LiClO_4$ in EC containing 4000 ppm H_2O. The water-driven hydrolysis rates were 0.5×10^{-12} and 10.9×10^{-12} mol/s at 40 and 60 °C, respectively.

$$H_2O + EC \rightarrow HO - CH_2 - CH_2 - OH + CO_2 \qquad (4.3)$$

The rates at 10 and 25 °C were too low to be evaluated. The apparent activation energies were calculated to be $\approx$133 kJ/mol from the data at 40 and 60 °C. Electrochemical oxidation of EC producing CO and CO_2 occurred at >4.5 V (vs. Li/Li$^+$) and was enhanced by the presence of 4000 ppm H_2O. The apparent activation energies were $\approx$104 and $\approx$92 kJ/mol for EC oxidation in dry and 4000 ppm water-containing electrolytes, respectively. The authors reported that in the presence of 4000 ppm H_2O, the onset of O_2 evolution (Eq. 4.4) was observed at $\approx$5.3 V versus Li/Li$^+$ on conductive carbon in cathodes.

$$2H_2O \rightarrow O_2 + 4H^+ + 4e^- \qquad (4.4)$$

H_2O in electrolytes is related to the oxidation of CB, producing CO and CO_2 (Eqs. 4.5 and 4.6).

$$C + H_2O \rightarrow CO + 2H^+ + 2e^- \qquad (4.5)$$

$$C + 2H_2O \rightarrow CO_2 + 4H^+ + 4e^- \qquad (4.6)$$

At 25 °C, carbon oxidation began when the potential reached approximately 4.8 V *vs.* Li/Li$^+$, even in the absence of added H_2O. The anodic oxidation rates of Super C65 carbon at 25 °C were 1.5×10^{-9} and 1.0×10^{-8} mol$_{Carbon}$/(s·g$_{Carbon}$) in the presence of <20 and 4000 ppm H_2O in EC with 2 M LiClO$_4$, respectively. The rate of oxidation of carbon was comparable to that of electrolyte oxidation. At 60 °C, the rate of electrolyte oxidation was higher than that of carbon oxidation. The apparent activation energies were $\approx$65 and $\approx$54 kJ/mol for carbon oxidation in dry and 4000 ppm water-containing electrolytes, respectively (Fig. 4.5). In addition, at 60 °C, $\approx$15 wt% carbon was oxidized at 5 V (vs. Li/Li$^+$) in only 100 h [28].

Fig. 4.5 Arrhenius plot for the anodic oxidation rates of Super C65 carbon and EC at a potential of 5.0 V (vs. Li/Li$^+$) for temperatures between 10 and 60 °C in both dry electrolyte (<20 ppm H_2O, open symbols) and water-containing electrolyte (4000 pp H_2O, solid symbols). The molar anodic oxidation rates are normalized to the mass of carbon and the electrolyte in EC with 2 M LiClO$_4$. [28]. Images reproduced with permission from the Electrochemical Society

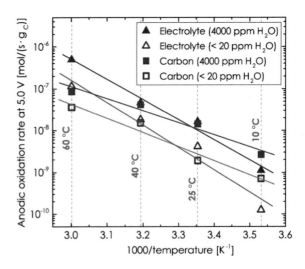

References

1. Yan C, Xu R, Xiao Y, Ding JF, Xu L, Li BQ, Huang JQ (2020) Adv Funct Mater 30:1909887
2. Heiskanen SK, Kim J, Lucht BL (2019) Joule 3:2322
3. Peled E, Menkin S (2017) J Electrochem Soc 164(7):A1703
4. An SJ, Li J, Daniel C, Mohanty D, Nagpure S, Wood DL III (2016) Carbon 105:52
5. Verma P, Maire1 P, Novák PA (2010) Electrochim Acta 55: 6332
6. Cheng CS, Wang FM, Rick J (2012) Int J Electrochem Sci 7:8676
7. Kitz PG, Novák P, Berg EJ (2020) ACS Appl Mater Interfaces 12: 15934
8. Cheng XB, Zhang R, Zhao CZ, Wei F, Zhang JG, Zhang Q (2016) Adv Sci 3:1500213
9. Cheng XB, Zhang R, Zhao CZ, Zhang Q (2017) Chem Rev 117:10403
10. Nie M, Abraham DP, Chen Y, Bose A, Lucht BL (2013) J Phys Chem C 117:13403
11. Gauthier M, Carney TJ, Grimaud A, Giordano L, Pour N, Chang HH, Fenning DP, Lux SF, Paschos O, Bauer C, Maglia F, Lupart S, Lamp P, Shao-Horn Y (2015) J Phys Chem Lett 6:4653
12. Saito M, Fujita M, Aoki Y, Yoshikawa M, Yasuda K, Ishigami R, Nakata Y (2016) Nucl Instr Meth B 371:273
13. Nazri G, Muller RH (1985) J Electrochem Soc 132:2050
14. Li Z, Xu F, Li C, Wang P, Yi W, Wang S, Yang L, Yan C, Li S (2021) ACS Appl Energy Mater 4:1199
15. Dedryvère R, Gireaud L, Grugeon S, Laruelle S, Tarascon JM, Gonbeau D (2005) J Phys Chem B 109:15868
16. Joho F, Rykart B, Imhof R, Novák P, Spahr M, Monnier A (1999) J Power Sources 81–82:243
17. Bernhard R, Metzger M, Gasteiger HA (2015) J Electrochem Soc 162(10):A1984
18. Jeschull F, Maibach J, Félix R, Wohlfahrt-Mehrens M, Edström K, Memm M, Brandell D (2018) ACS Appl Energy Mater 1: 5176
19. Jeschull F, Brandell D, Wohlfahrt-Mehrens M, Memm M (2017) Energy Technol 5:2108
20. Edström K, Gustafsson T, Thomas JO (2004) Electrochim Acta 50:397
21. Kim K, Ma H, Park S, Choi NS (2020) ACS Energy Lett 5:1537
22. Wang H, Li X, Li F, Liu X, Yang S, Ma J (2021) Electrochem Commun 122: 106870
23. Zhao D, Li S (2020) Front Chem 8:821
24. Wagner R, Korth M, Streipert B, Kasnatscheew J, Gallus DR, Brox S, Amereller M, Cekic-Laskovic I, Winter M (2016) ACS Appl Mater Interfaces 8:30871
25. Yang L, Ravdel B, Lucht BL (2010) Electrochem Solid-State Lett 13:A95
26. Wu J, Membreno N, Yu WY, Wiggins-Camacho JD, Flaherty DW, Mullins CB, Stevenson KJ (2012) J Phys Chem C 116:21208
27. Bernardine LD, Rinkel DS, Hall IT, Clare PG (2020) J Am Chem Soc 142:15058
28. Metzger M, Marino C, Sicklinger J, Haering D, Gasteiger HA (2015) J Electrochem Soc 162(7):A1123

Chapter 5
Cathode and Anode Preparation by the Aqueous Process

Abstract Recently, water-soluble and aqueous polymers (water-based polymers) have attracted much attention as binders for LIBs because of the need for low-cost materials and environmentally compatible electrode fabrication processes. The water process for the fabrication of cathodes and anodes still has the problems that should be resolved. In this chapter, the origin of the problems, the solutions and the performance of cathodes and anodes that are treated with the processes and techniques is summarized.

Keywords Water-based polymer binder · Water-based slurry · Corrosion · Current collector · Coating

5.1 Water Fabrication Process of Cathode and Anode

Recently, aqueous processes for the fabrication of anode and cathode electrodes in LIBs with water-soluble and aqueous polymers, i.e., water-based polymers, have attracted much attention as binders for use in environmentally compatible LIB fabrication processes and for reducing the cost of LIBs [1–7]. Moreover, conventional PVdF polymer binder is dissolved in NMP, which is listed as a carcinogenic chemical with reproductive toxicity [8, 9]. To maintain the battery performance that cathodes prepared with conventional PVdF binders exhibit, replacing PVdF with water-based polymer binders is not always easy due to, e.g., the low resistance of water-based polymer binders to electro-oxidation and the low water resistance of cathode materials, as well as the difficulty associated with the uniform coating of water-based slurries (containing water-based polymer binders, cathode materials, conducting carbon, etc.) on aluminum (Al) current collectors [10]. When cathode materials contact H_2O in the preparation process of aqueous slurry containing cathode material, conductive additive and binder, Li^+ ions leach out from the cathode material surfaces. To balance the charge in cathode materials, cation exchange of H^+/Li^+ occurs. That is, Li^+ ions migrate away from the cathode materials, and H^+ migrates toward the cathode materials. As a result, the pH of the aqueous slurry increases. If the slurries are coated on Al current collectors, the surface of Al is corroded (Eq. 5.1), cathode material layers formed on the Al current collectors fall off from the Al current collectors and the cathode loss capacity rapidly drops [11, 12].

$$Al + H_2O + OH^- \rightarrow AlO_2^- + 3/2 \, H_2 \uparrow \tag{5.1}$$

Stainless steel foils with anticorrosion properties in strong alkaline slurries have been developed by the Japanese company *Nippon Steel* [13]. When water-based slurries are cast on the current collector surface in the fabrication processes of anodes and cathodes, the water-based slurries are impacted by the current collector surfaces because H_2O cannot contact metal surfaces well. To inhibit the increase in pH of the slurry, surface coating of cathode materials [14–16] and notarization of the slurry with acids [17–19] and carbon dioxide gas [20] have been proposed.

Although cathode materials are sensitive to moisture, as mentioned above, there are many examinations in which nontreated cathode materials are added to aqueous slurries and cathodes prepared with nontreated cathode materials, and water-based binders exhibit relatively stable cathode performance. The results are summarized in Table 5.1 [21–31].

5.2   Weakness of Cathode Materials Against H_2O

As mentioned above, cathode materials showed a weak defense against moisture. However, cathodes prepared with an aqueous slurry in which the cathode materials were immersed in H_2O during the preparation process of slurries exhibited relatively stable charging/discharging cycle performance. The authors considered that among the experiments described in this book, the preparation of slurries of cathode materials and casting of the slurries consumed a short period and the cathode materials were not seriously damaged by H_2O. Conversely, in several papers, contact with H_2O for a long period was applied to cathode materials. The test conditions for examining the degree of waterproofing of cathode materials should be unified. Hawley et al. examined commercially available waterproof cathode materials of $LiCoO_2$, LFP, $LiMn_2O_4$, $LiNi_{0.80}Co_{0.15}Al_{0.05}O_2$ and $LiNi_{0.5}Mn_{0.3}Co_{0.2}O_2$ after three days of H_2O exposure [32]. $LiNi_{0.80}Co_{0.15}Al_{0.05}O_2$ and $LiNi_{0.5}Mn_{0.3}Co_{0.2}O_2$ exhibited significantly greater pH values (11.5–12.5) than Ni-free $LiCoO_2$, LFP and $LiMn_2O_4$ (9.0–10.5). The surface composition of Li^+ largely decreased on $LiNi_{0.80}Co_{0.15}Al_{0.05}O_2$ and $LiNi_{0.5}Mn_{0.3}Co_{0.2}O_2$ due to much more Li^+ dissolution from the surface into H_2O (Fig. 5.1). When cathodes were prepared with water-treated $LiCoO_2$, LFP, $LiMn_2O_4$, $LiNi_{0.80}Co_{0.15}Al_{0.05}O_2$ and $LiNi_{0.5}Mn_{0.3}Co_{0.2}O_2$ and high-molecular weight polyacrylic acid (PAA), PAA was able to modify the pH and provide adequate binding to the current collector and could contribute to stable charging/discharging capacities even on $LiNi_{0.80}Co_{0.15}Al_{0.05}O_2$ and $LiNi_{0.5}Mn_{0.3}Co_{0.2}O_2$ [32].

Table 5.1 Summary of the performance of the cathode prepared with water-based slurries

Cathode materials	Synthesis condition of cathode materials	Slurry preparation condition				Capacity fading				References
		Concentration of active material in water	pH of water in which active materials are immersed	Period of contacting with water	Water-based binder	Cutoff voltage (V vs. Li/Li+)	C-rate (C)	Cycle number	Capacity retention[1) (%)	
LiFePO4	Citric sol–gel method	50 g L^{-1}	7–8	24 h	–	3.1–3.8	0.05	2	97	[21]
Carbon-coated LiFePO4	Phostech, Canada	A slurry consisting of 90 wt.% of LiFePO4, 6 wt.% carbon black, 2 wt.% CMC and 2 wt.% WSB	–	–	Water-soluble elastomer (WSB) from ZEON Corp, Japan	2.5–4	1	200	93	[22]
LiFePO4	Hydrothermal method with starting materials LiOH·H2O, H3PO4 and FeSO4 · 7H2O	LiFePO4:super P carbon black: chitosan = 90:5:5 (wt.%) aqueous slurry was prepared with deionized water containing 0.5% acetic acid (wt.%) as a solvent for blending	–	12 h	Chitosan	2.7–4.3	1	100	90	[23]
LiCoO2	Umicore	50 g L^{-1} in water	12	>16 h	–	3–4.2	1	100	74	[24]
LiCoO2	–	LCO, AB, SBRx, and CMC were thoroughly mixed at a ratio of 80:10:0.5:1.5 (m/m) with distilled water	–	–	SBRx latex (NIPPON A & L, INC.)	3–4.5	0.1	50	93	[25]
LiNi0.5Mn1.5O4 (LNMO)	Coprecipitation method	LNMO: conductive carbon: binder = 87:10:3 (wt.%)	–	2	Natural guar gum	3.5–4.9	1	400	92	[26]

(continued)

Table 5.1 (continued)

Cathode materials	Synthesis condition of cathode materials	Slurry preparation condition				Capacity fading				References
		Concentration of active material in water	pH of water in which active materials are immersed	Period of contacting with water	Water-based binder	Cutoff voltage (V vs. Li$^+$)	C-rate (C)	Cycle number	Capacity retention[1]) (%)	
LiNi$_{1/3}$Mn$_{1/3}$Co$_{1/3}$O$_2$/graphite full cell	TODA, average particle size: d$_{90}$ = 10 μm SLP 30 graphite, TIMCAL, average particle size: d$_{90}$ = 32 μm	NMC: super C45: CMC = 88: 7: 5 (wt.%) Graphite: super C45: CMC = 90: 5: 5 (wt.%)	–	3	Sodium carboxymethyl cellulose (CMC, Dow Wolff Cellulosics, Walocel CRT 2000 PPA 12)	3–4.3	1	2000	70	[27]
LiNi$_{0.4}$Co$_{0.2}$Mn$_{0.4}$O$_2$ (NCM)	Coprecipitation method	NCM: super C45: CMC = 80: 15: 5 (wt.%)	–	2	CMC (Dow Wolff Cellulosics, Walocel CRT 2000 PPA 12)	3–4.3	2	200	92	[28]
LiNi$_{0.5}$Mn$_{0.3}$Co$_{0.2}$O$_2$ (NMC532)	Umicore; average particle size: d$_{50}$ = 12 μm	NMC532: super C65: CMC: latex: H$_3$PO$_4$ = 91.1: 4: 1: 3: 0.9 (wt.%)	11.8	–	Latex binder (Zeon)	3–4.3	0.5	250	85	[29]
LiNi$_{0.8}$Mn$_{0.1}$Co$_{0.1}$O$_2$ (NMC 811)	Targray	NMC811: carbon black (Denka Li-100): CMC, (Ashland):Ashland acrylic emulsion = 90: 5:4:1 (wt.%)	12	1 week	CMC (Ashland)	2.6–4.5	0.3	500	84	[27]
LiNi$_{0.8}$Co$_{0.15}$Al$_{0.05}$O$_2$	NCA, NAT-1050, TODA	NCA:water = 1:3 (wt.%) NCA in water was then filtered; the resulting NCA powder was filtrated and dried at 110 °C for 16 h	–	8 h	–	3–4.3	0.1	50	81	[30]

(continued)

Table 5.1 (continued)

Cathode materials	Synthesis condition of cathode materials	Slurry preparation condition				Capacity fading				References
		Concentration of active material in water	pH of water in which active materials are immersed	Period of contacting with water	Water-based binder	Cutoff voltage (V vs. Li/Li$^+$)	C-rate (C)	Cycle number	Capacity retention[1]) (%)	
Li[Li$_{0.2}$Mn$_{0.56}$Ni$_{0.16}$Co$_{0.08}$]O$_2$	Solid-state reaction method from lithium hydroxide hydrate and manganese–nickel–cobalt hydroxide precursors	Li[Li$_{0.2}$Mn$_{0.56}$Ni$_{0.16}$Co$_{0.08}$]O$_2$: super P (TIMCAL):CMC = 85: 10: 5 (wt%)	9–11	3 h	CMC (Dow Wolff cellulosics, Walocel CRT 2000 PPA 12)	2.5–4.8	1	50	83	[31]

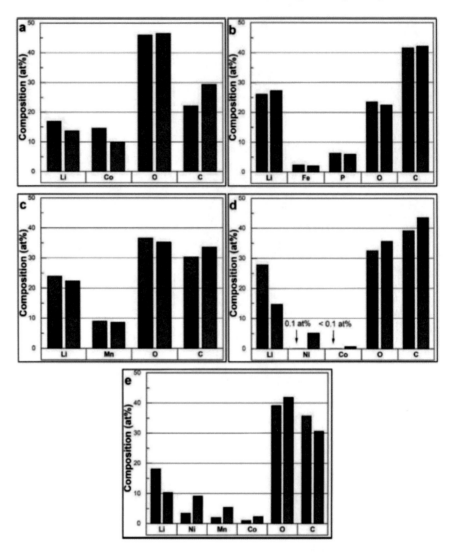

Fig. 5.1 Surface composition (in at%) of the dry (red) and wet (navy) **a** $LiCoO_2$, **b** $LiFePO_4$, **c** $LiMn_2O_4$, **d** $LiNi_{0.80}Co_{0.15}Al_{0.05}O_2$ and **e** $LiNi_{0.5}Mn_{0.3}Co_{0.2}O_2$ samples according to XPS peak fitting [32]. Reprinted from J. Power Sources, 466, Hawley, W.B.; Parejiya, A.; Bai, Y.; Meyer III, H.M.; Wood III, D.L. Li, J.: Lithium and transition metal dissolution due to aqueous processing in lithium-ion battery cathode active materials, 228315, Copyright (2020), with permission from Elsevier

5.3 Property of Other Anode Materials Against H_2O

In anodes, the aqueous fabrication process of graphite has already been developed in some commercially available Li-ion batteries [1, 33, 34]. SBR and CMC-Na are representative water-based binders [35, 36]. Aqueous processes of anodes have also been applied to high-capacity anode materials of MCMB [37], Si [38–40], Sn [41], TiO_2 [42] and $ZnFe_2O_4$ [43]. These anode materials can be applied to aqueous processes potentially because they do not contain Li^+ ions. Among high-capacity anode materials, LTO contains Li^+ ions and does not lose performance after contact with H_2O [44, 45]. In the aqueous process, LTO absorbs H_2O and CO_2 and reacts with H_2O and CO_2 to form Li_2CO_3 on the surface of LTO. Li_2CO_3, which inhibits lithium intercalation/deintercalation, is formed on the surface of LTO. Because Li_2CO_3 can be removed by heat treatment, the effect of the presence of Li_2CO_3 on the electrochemical performance is negligible in LTO [46].

References

1. Bresser D, Buchholz D, Moretti A, Varzi A, Passerini S (2018) Energy Environ Sci 11:3096
2. Iturrondobeitia A, Kvasha A, Lopez del Amo JM, Colin JF, Sotta D, Urdampilleta I, Casas-Cabanas M (2017) Electrochim Acta 247: 420
3. Chou WY, Jin YC, Duh JG, Lu CZ, Liao SC (2015) Appl Surf Sci 355:1272
4. Chou SL, Pan Y, Wang JZ, Liu HK, Dou SX (2014) Phys Chem Chem Phys 16:20347
5. Du Z, Rollag KM, Li J, An SJ, Wood M, Sheng Y, Mukherjee PP, Daniel C, Wood DL III (2017) J Power Sources 354:200
6. Valvo M, Liivat A, Eriksson H, Tai CW, Edström K (2017) Chemsuschem 10:2431
7. Li JT, Wu ZY, Lu YQ, Zhou Y, Huang AS, Huang L, Sun SG (2017) Adv Energy Mater 7:1701185
8. Wood DL III, Li J, Daniel C (2015) J Power Sources 275:234
9. Solomon GM, Morse EP, Garbo MJ, Milton DK (1996) J Occup Environ Med 38:705
10. Wakao T, Gunji T, Jeevagan AJ, Mochizuki Y, Kaneko S, Baba K, Watanabe M, Kanda Y, Murakami K, Omura M, Kobayashi G, Matsumoto F (2014) ECS Trans 58(25):19
11. Church BC, Kaminski DT, Jiang J (2014) J Mater Sci 49:3234
12. Li SY, Church BC (2016) Mater Corros 67:978
13. Unno H, Nagata T, Fujimoto N, Fukuda M (2019) Nippon Steel Gihou 412:173
14. Tanabe T, Liu YB, Miyamoto K, Irii Y, Maki F, Gunji T, Kaneko S, Ugawa S, Lee H, Ohsaka T, Matsumoto F (2017) Electrochim Acta 258:1348
15. Notake K, Gunji T, Kokubun H, Kosemura S, Mochizuki Y, Tanabe T, Kaneko S, Ugawa S, Lee H, Matsumoto F (2016) J Appl Electrochem 46:267
16. Watanabe T, Hirai K, Ando F, Kurosumi S, Ugawa S, Lee H, Irii Y, Maki F, Gunji T, Wu J, Ohsaka T, Matsumoto F (2020) RSC Adv 10:13642
17. Wood M, Li J, Ruther RE, Du Z, Self EC, Meyer HM III, Daniel C, Belharouak I, Wood DL III (2020) Energy Stor Mater 24:188
18. Bauer W, Çetinel FA, Müller M, Kaufmann U (2019) Electrochim Acta 317:112
19. Kuenzel M, Bresser D, Diemant T, Carvalho DV, Kim GT, Behm RJ, Passerini S (2018) Chemsuschem 11:562
20. Kimura K, Sakamoto T, Mukai T, Ikeuchi Y, Yamashita N, Onishi K, Asami K, Yanagida M (2018) J Electrochem Soc 165:A16

21. Porcher W, Moreau P, Lestriez B, Jouanneau S, Guyomard D (2008) Electrochem Solid-State Lett 11:A4
22. Guerfi A, Kaneko M, Petitclerc M, Mori M, Zaghib K (2007) J Power Sources 163:1047
23. Prasanna K, Subburaj T, Jo YN, Lee WJ, Lee CW (2015) ACS Appl Mater Interfaces 7:7884
24. Martínez CAP, Exantus C, Dallel D, Alié C, Calberg C, Liquet D, Eskenazi D, Deschamps F, Job N, Heinrichs B (2019) Adv Mater Technol 4:1900499
25. Isozumi H, Horiba T, Kubota K, Hida K, Matsuyama T, Yasuno S, Komaba S (2020) J Power Sources 468: 228332
26. Kuenzel M, Choi H, Wu F, Kazzazi A, Axmann P, Wohlfahrt-Mehrens M, Bresser D, Passerini S (2020) Chemsuschem 13:2650
27. Loeffler N, Zamory JV, Laszczynski N, Doberdo I, Kim GT, Passerini S (2014) J Power Sources 248:915
28. Chen Z, Kim GT, Chao D, Loeffler N, Copley M, Lin J, Shen Z, Passerini S (2017) J Power Sources 372:180
29. Bichon M, Sotta D, Dupré N, Vito ED, Boulineau A, Porcher W, Lestriez B (2019) ACS Appl Mater Interfaces 11:18331
30. Hofmann M, Kapuschinski M, Guntow U, Giffin GA (2020) J Electrochem Soc 167: 140535
31. Li J, Klopsch R, Nowak S, Kunze M, Winter M, Passerini S (2011) J Power Sources 196:7687
32. Hawley WB, Parejiya A, Bai Y, Meyer III HM, Wood III DL, Li J (2020) J Power Sources 466: 228315
33. Cholewinski A, Si P, Uceda M, Pope M, Zhao B (2021) Polymers 13:631
34. Asenbauer J, Eisenmann T, Kuenzel M, Kazzazi A, Chen Z, Bresser D (2020) Energy Fuels 4:5387
35. Buqa H, Holzapfel M, Krumeich F, Veit C, Novák P (2006) J Power Sources 161:617
36. Lee JH, Lee S, Paik U, Choi YM (2005) J Power Sources 147:249
37. Courtel FM, Niketic S, Duguay D, Abu-Lebdeh Y, Davidson IJ (2011) J Power Sources 196:2128
38. Beattie SD, Larcher D, Morcrette M, Simon B, Tarascon JM (2008) J Electrochem Soc 155:A158
39. Magasinski A, Zdyrko B, Kovalenko I, Hertzberg B, Burtovyy R, Huebner CF, Fuller TF, Luzinov I, Yushin G (2010) ACS Appl Mater Interfaces 2:3004
40. Klamor S, Schröder M, Brunklaus G, Niehoff P, Berkemeier F, Schappacher FM, Wintera M (2015) Phys Chem Chem Phys 17:5632
41. Zhao Y, Yang L, Liu D, Hu J, Han L, Wang Z, Pan F (2018) ACS Appl Mater Interfaces 10:1672
42. Mancinia M, Nobilib F, Tossici R, Marassi R (2012) Electrochim Acta 85:566
43. Zhang R, Yang X, Zhang D, Qiu H, Fu Q, Na H, Guo Z, Du F, Chen G, Wei Y (2015) J Power Sources 285:227
44. Carvalho DV, Loeffler N, Kim GT, Marinaro M, Wohlfahrt-Mehrens M, Passerini S (2016) Polymers 8:276
45. Pohjalainen E, Räsänen S, Jokinen M, Yliniemi K, Worsley DA, Kuusivaara J, Juurikivi J, Ekqvist R, Kallio T, Karppinen M (2013) J Power Sources 226:134
46. Gao Y, Wang Z, Chen L (2014) J Power Sources 245:684

Chapter 6
Aqueous Battery

Abstract From the viewpoints of safety, low-cost and environmentally friendly products, the development of reversible aqueous batteries is expected to yield very significant results. However, conventional materials such as cathodes and anodes are weak against H_2O. In order to overcome weakness of the materials, various ideas are proposed and examined. Recent studies are reviewed in this chapter.

Keywords Reversible aqueous battery · Li metal · Potential window · Surface coating · Water-in-salt electrolytes

6.1 Application of Li Metal Anode to RABs

RABs have been widely investigated for large-scale energy storage devices in view of their high safety and low cost. The principal disadvantage is the limited thermo-dynamic electrochemical window of H_2O [1]. The achievement of a wide potential window in aqueous electrolytes is currently being challenged [2], for example, through the use of highly concentrated electrolytes: water-in-salt [3, 4]. Many review papers on resolving the problem of the potential window have been published [5–7]. In this section, as mentioned above, the reactions of H_2O with materials in RABs will be the focus of the review.

Employment of Li metal as an anode in RABs is attractive because of its high capacity (3861 mAh g^{-1}) for an anode. However, Li metal is seriously vulnerable to H_2O. To apply Li metal anodes to RABs, various surface-coated Li metal anodes have been developed. For example, Wang et al. proposed a Li metal anode coated with a sandwiched polymer membrane of PVdF/polymethyl methacrylate/PVdF saturated with 1 mol L^{-1} LiClO$_4$ solution in EC and a γ-Li$_3$PO$_4$-type structure (LISICON) film consisting of Li$_2$O–Al$_2$O$_3$–SiO$_2$–P$_2$O$_5$–TiO$_2$–GeO$_2$. The coated layer formed on Li metal showed high Li$^+$ conductivity and prohibited H$_2$ evaluation on Li metal (Fig. 6.1) [8]. In the RABs, the coated lithium metal anode and LiMn$_2$O$_4$ cathode were dipped into a 0.5 mol L^{-1} Li$_2$SO$_4$ aqueous electrolyte. The average discharge voltages were approximately 4.0 V. After 30 full cycles, the discharge capacity remained very stable at 115 mAh g^{-1}, which indicated that no obvious capacity loss occurred in the first 30 cycles. It has been proposed that Li$^+$ ion-conducting glass–ceramic

© The Author(s), under exclusive license to Springer Nature Singapore Pte Ltd. 2022   49
F. Matsumoto and T. Gunji, *Water in Lithium-Ion Batteries*, SpringerBriefs in Energy,
https://doi.org/10.1007/978-981-16-8786-0_6

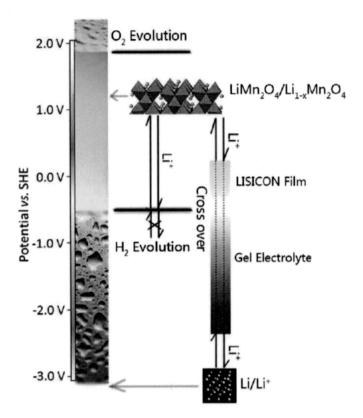

Fig. 6.1 Schematic illustration of the potential of Li⁺ ions during the movement between LiMn₂O₄ in the aqueous electrolyte and the coated lithium metal. Reproduced with permission [8]. Copyright 2013, Nature Publishing Group

plates that do not exhibit H_2O permeation separate organic electrolytes for Li anodes and aqueous electrolytes for cathodes [9]. This idea is applicable to aqueous Li-air batteries [10–12]. Another example demonstrates the application of Li⁺ ion-conducting ceramic coating to silicon anodes [13].

6.2   Application of Other Anodes to RABs

Graphite and silicon anodes are situated far beyond the limit of the potential window of H_2O. $LiTi_2(PO_4)_3$ [14], TiP_2O_7 [15], LiV_2O_8 [16] and VO_2 [17] anodes fit in the potential window of H_2O (Fig. 6.2) [2]. Li⁺ intercalation and deintercalation on graphite anodes were realized in water-in-salt electrolytes. In high-concentration LiTFSI aqueous electrolytes, LiTFSI dominated in inner-Helmholtz interfacial regions of the anode surface, while H_2O molecules were excluded from

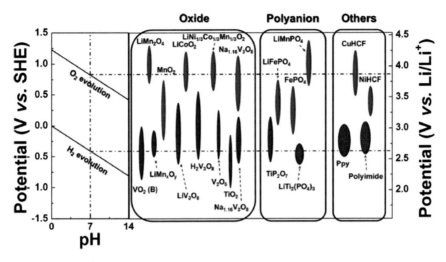

Fig. 6.2 Electrode materials for RABs. Electrode materials were categorized as oxide compounds, polyanionic compounds and other compounds. Reprinted with permission from Ref. [2]. Copyright 2014 American Chemical Society

the anode surface. During initial charging, TFSI⁻ anions were reduced preferentially to form a dense SEI on the anode surface. Because the SEI layer allows ionic conduction but prohibits electronic conduction, the SEI layer protects the graphite anode from hydrogen evolution, and SEI-covered graphite can expand the potential window for Li⁺ intercalation/deintercalation on graphite [18]. The same discussion has been reported in the case of a Mo_6S_8 anode in a water-in-salt electrolyte [19]. In addition, Suo et al. discussed the mechanism underlying SEI formation in aqueous electrolytes from TFSI⁻ ions [20]. The performance of silicon anodes in water-in-salt electrolytes has not yet been reported.

6.3 Which Cathode Materials Can Be Applied to RABs

Spinel $LiMn_2O_4$ is a cathode material that is quite often examined for RABs because the Li⁺ intercalation/deintercalation potential is in the potential window of conventional aqueous electrolytes [21]. As other examples, the performance of conventional cathode materials observed in conventional aqueous electrolytes and concentrated aqueous electrolytes is summarized in Table 6.1 [21–28]. Recently, hydrate melt electrolytes, such as $Li(PTFSI)_{0.6}(TFSI)_{0.4} \cdot 1.15H_2O$ (LiPTFSI: $LiN(SO_2CF_3)(SO_2C_2F_5)$), have also been developed to expand the potential windows to 5 V [29]. The potential windows of water-in-salt electrolytes and the battery performance of water-in-salt-based batteries have been reviewed by Bélanger [30] and Tarascon [31], respectively.

Table 6.1 Summary of cathode performance in aqueous LIBs

Cathode materials	Conditions of aqueous electrolyte	Cycle performance						References
		Cutoff voltage (V vs. SCE)	C-rate (C)	Discharge capacity at first cycle (mAh g^{-1})	Capacity retention[1] (%)	Cycle number for estimating capacity retention	Average voltage for intercalation/deintercalation of Li$^+$ on cathode materials (V)	
Li$_{1.06}$Mn$_2$O$_4$	1 M Li$_2$SO$_4$	3.4–4.4	1	110	96	50	4.1	[21]
LiMn$_2$O$_4$	9–15 M LiTFSI	4.5–5.6	1	106	82	1400	5.2	[22]
Fd-3 m LiNi$_{0.5}$Mn$_{1.5}$O$_4$	21 M LiTFSI (pH 5)	3.2–52–5.05	0.5	125	0.075%/ cycle at 5 C	–	4.5	[23]
Li$_{1.05}$Cr$_{0.10}$Mn$_{1.85}$O$_4$	9 M LiNO$_3$ + 1wt% vinylene carbonate	3–3.87	0.05	112	82	100	3.75	[24]
LiCoO$_2$	5 M LiNO$_3$	3.7–4.3	1	115	87	90	4.2	[25]
LiCoO$_2$	0.5 M Li$_2$SO$_4$	3.8–4.05 (cyclic voltammetry)	-	-	-	-	3.92	[26]
LiCoO$_2$	1 M lithium bis(fluorosulfonyl)imide (LiFSI)	3.1–4.8	0.1	52.53	67.8	500	4.3	[27]
LiCoO$_2$	1 M lithium sulfonylbis(fluorosulfonyl)imide (LiSFSI)	3.1–4.8	0.1	113.50	97.3	500	4.3	[27]
LiCoO$_2$	1 M lithium carbonylbis(fluorosulfonyl)imide (LiCFSI)	3.1–4.8	0.1	95.0	89.6	500	4.3	[27]
LiFePO$_4$	0.5 M Li$_2$SO$_4$	3.7–4.7	5	130	67	10	4.1	[28]

Similar to the SEI formed on the anode surface, the CEI also contributes to expansion of the potential windows in the direction of higher potential [32–34]. The dissolution of cathode materials in aqueous electrolytes is a serious problem in cathode performance, as discussed in the previous section.

Although a report on the dissolution of cathode materials for LIBs in water-in-salt cannot yet be found, water-in-salt electrolytes suppress the cathode dissolution of $Na_3V_2(PO_4)_3$ cathodes for sodium ion batteries [35]. Many studies have reported the degradation of cathode performance and full cell performance. However, the reason for the degradation of performance, that is, the degradation results from the potential window or dissolution of cathode materials in aqueous electrolytes, is vague and obscure. The amount of metal ions dissolved from cathode materials during the charging/discharging process and open-circuit conditions should be analyzed carefully to clarify the underlying reason. The CEI layer can also suppress the dissolution of the cathode surface in $Zn_3V_2O_7(OH)_2 \cdot 2H_2O$ for zinc ion batteries [36]. A surface-coated LFP cathode coated with AlF_3 beforehand was also applied to a 1 M Li_2SO_4 aqueous electrolyte [37].

1. Capacity retention = (discharged capacity observed with air-exposed cathode materials)/(discharged capacity observed with cathode materials after storage without atmospheric exposure) × 100 (%)

As mentioned above, protons can be co-intercalated into the layer of cathode materials parallel to the intercalation of Li^+ ions. The intercalation of protons affects the cathode performance. The intercalation of protons into the layered structure of cathode materials was shown to cause a distortion of the local structure of the cathode materials and reduced lithiation efficiency [38]. In addition, protons intercalated into the layered structures bonded covalently to the oxygen ions in the frameworks and solidified barriers to the diffusion of Li^+ ions in the layered structures [39]. The degree of proton intercalation depends on the type of cathode material and the pH of the aqueous electrolyte. It has been reported that proton intercalation does not occur in $LiMn_2O_4$ and LFP [1]. Xia et al. reported that in charge/discharge tests in 1 M Li_2SO_4, $LiMn_2O_4$ showed stable Li^+ intercalation in the solution at pH 7, $LiCo_{1/3}Ni_{1/3}Mn_{1/3}O_2$ at pH 11 and $LiCoO_2$ at pH 9, indicating that proton intercalation did not predominantly occur at the critical pH of aqueous electrolytes [40, 41]. Gu et al. examined the proton-induced dysfunction mechanism of $LiCo_{1/3}Ni_{1/3}Mn_{1/3}O_2$ and $LiMn_2O_4$ cathodes in RABs in experimental and density functional theory (DFT) theoretical research [42]. Based on the concentration of H^+ and Li^+ in aqueous electrolytes, the critical pH values of $LiCo_{1/3}Ni_{1/3}Mn_{1/3}O_2$ and $LiMn_2O_4$ cathodes for normal function in aqueous electrolytes could be found. The diagram of pH vs. Li^+ concentration showing the critical conditions for normal function of the $LiCo_{1/3}Ni_{1/3}Mn_{1/3}O_2$ and $LiMn_2O_4$ cathodes was calculated in that paper. The results showed stable Li^+ intercalation of $LiCo_{1/3}Ni_{1/3}Mn_{1/3}O_2$ in the solution at pH 7. Conversely, LFP exhibited considerable performance degradation when the pH of the aqueous electrolytes exceeded 11 due to the loss of Li, Fe and P dissolution during prolonged charge–discharge in aqueous medium [43]. In $LiNiO_2$, which is susceptible to damage in aqueous electrolytes, as mentioned above, proton intercalation formed a $NiOOH_x$ film on the $LiNiO_2$ surface during the initial electrochemical cycles. The cycle performance of $LiNiO_2$ in aqueous electrolytes was extremely poor. XPS and transmission electron microscopy (TEM) analysis confirmed that the formation of a spinel-like Ni_3O_4 species that was electrochemically inactive at the subsurface was caused by the absence of Li in the $NiOOH_x$ film at the surface of $LiNiO_2$, which was the reason for the low performance of the $LiNO_2$ cathode in aqueous electrolytes [44]. Additionally, in this case to our knowledge, the relationship between the degree of proton intercalation and the concentration/type of Li salts in aqueous electrolytes unfortunately could not be found.

References

1. Wang Y, Yi J, Xia Y (2012) Adv Energy Mater 2:830
2. Kim H, Hong J, Park KY, Kim H, Kim SW, Kang K (2014) Chem Rev 114:11788
3. Suo L, Borodin O, Gao T, Olguin M, Ho J, Fan X, Luo C, Wang C, Xu K (2015) Science 350:938
4. Li M, Wang C, Chen Z, Xu K, Lu J (2020) Chem Rev 120:6783
5. Demir-Cakan R, Palacin MR, Croguennec L (2019) J Mater Chem A 7:20519
6. Wang H, Tan R, Yang Z, Feng Y, Duan X, Ma J (2020) Adv Energy Mater 2000962.
7. Huang J, Guo Z, Ma Y, Bin D, Wang Y, Xia Y (2019) Small Methods 3:1800272
8. Wang X, Hou Y, Zhu Y, Wu Y, Holze R (2013) Sci Rep 3:1401
9. Zhang M, Huang Z, Shen Z, Gong Y, Chi B, Pu J, Li J (2017) Adv Energy Mater 7:1700155
10. Lu J, Li L, Park JB, Sun YK, Wu F, Amine K (2014) Chem Rev 114:5611
11. Kim JK, Yang W, Salim J, Ma C, Sun C, Li J, Kim Y (2014) J Electrochem Soc 16:A285
12. Bai1 F, Kakimoto K, Shang X, Mori D, Taminato S, Matsumoto M, Takeda Y, Yamamoto O, Izumi H, Minami H, Imanishi N (2020) Front Energy Res. 8: 187
13. Paravannoor A, Panoth D, Praveen P (2020) Appl Sci 2:1831
14. Luo JY, Xia YY (2007) Adv Funct Mater 17:3877
15. Wen Y, Chen L, Pang Y, Guo Z, Bin D, Wang YG, Wang C, Xia Y (2017) ACS Appl Mater Interfaces 9:8075
16. Wang GJ, Zhao NH, Yang LC, Wu YP, Wu HQ, Holze R (2007) Electrochim Acta 52:4911
17. Zhang H, Cao D, Bai X (2019) J Power Sources 444: 227299
18. Yang C, Chen J, Qing T, Fan X, Sun W, Cresce AV, Ding MS, Borodin O, Vatamanu J, Schroeder MA, Eidson N, Wang C, Xu K (2017) Joule 1:122
19. Suo L, Borodin O, Gao T, Olguin M, Ho J, Fan X, Luo C, Wang C, Xu K (2015) Science 350:6265
20. Suo: L, Oh D, Lin Y, Zhuo Z, Borodin O, Gao T, Wang F, Kushima A, Wang Z, Kim HC, Qi Y, Yang W, Pan F, Li J, Xu K, Wang C (2017) J Am Chem Soc 139: 18670
21. Wang Y, Chen L, Wang Y, Xia Y (2015) Electrochim Acta 173:178
22. Wen Y, Ma C, Chen H, Zhang H, Li M, Zao P, Qiu J, Ming H, Cao G, Tang G (2020) Electrochim Acta 362: 137079
23. Wang F, Suo L, Liang Y, Yang C, Han F, Gao T, Sun W, Wang C (2017) Adv Energy Mater 7:1600922
24. Stojković IB, Cvjetićanin ND, Mentus SV (2010) Electrochem Commun 12:371
25. Ruffo R, Wessells C, Huggins RA, Cui Y (2009) Electrochem Commun 11:247
26. Levin EE, Vassiliev SY, Nikitina VA (2017) Electrochim Acta 228:114
27. Ahmed F, Rahman MM, Sutradhar SC, Lopa NS, Ryu T, Yoon S, Choi I, Lee S, Kim W (2019) Electrochim Acta 298:709
28. He P, Liu JL, Cui WJ, Luo JY, Xia YY (2011) Electrochim Acta 56:2351

29. Ko S, Yamada Y, Miyazaki K, Shimada T, Watanabe E, Tateyama Y, Kamiya T, Honda T, Akikusa J, Yamada A (2019) Electrochem Commun 104:106488
30. Amiri M, Bélanger D (2021) Chemsuschem 14:2487
31. Droguet L, Grimaud A, Fontaine O, Tarascon JM (2020) Adv Energy Mater 10:2002440
32. Lee C, Yokoyama Y, Kondo Y, Miyahara Y, Abe T, Miyazaki K (2021) Adv Energy Mater 11:2100756
33. Droguet L, Hobold GM, Lagadec MF, Guo R, Lethien C, Hallot M, Fontaine O, Tarascon JM, Gallant BM, Grimaud A (2021) ACS Energy Lett 6:2575
34. Liu S, Liu D, Wang S, Cai X, Qian K, Kang F, Li B (2019) J. Mater. Chem. A 7:12993
35. Yue J, Lin L, Jiang L, Zhang Q, Tong Y, Suo L, Hu YS, Li H, Huang X, Chen L (2020) Adv Energy Mater 10:2000665
36. Guo J, Ming J, Lei Y, Zhang W, Xia C, Cui Y, Alshareef HN (2019) ACS Energy Lett 4:2776
37. Tron A, Jo YN, Oh SH, Park YD, Mun J (2017) ACS Appl Mater Interfaces 9:12391
38. Oh H, Yamagishi H, Ohta T, Byon HR (2012) Mater Chem Front 5:3657
39. Gu X, Liu JL, Yang JH, Xiang HJ, Gong XG, Xia YY (2011) J Phys Chem C 115:12672
40. Wang YG, Luo JY, Wang CX, Xia YY (2006) J Electrochem Soc 153:A1425
41. Wang YG, Lou JY, Wu W, Wang CX, Xia YY (2007) J Electrochem Soc 154:A228
42. Shu Q, Chen L, Xia Y, Gong X, Gu X (2013) J Phys Chem C 117:6929
43. Yin Y, Wen YH, Lu YL, Cheng J, Cao GP, Yang YS (2015) Chin J Chem Phys 28:315
44. Lee C, Yokoyama Y, Kondo Y, Miyahara Y, Abe T, Miyazaki K (2020) ACS Appl Mater Interfaces 12:56076

Chapter 7
All-Solid-State LIBs

Abstract Recently, owing to safety issues, geometric restrictions related to the electrolyte and inhibition of the formation of Li dendrites in LIBs, solid-state electrolytes have been extensively examined. The development of high-power all-solid-state batteries using sulfide- and oxide-based solid electrolytes has been reported. The possibility of realizing high-power all-solid-state batteries by 2030 has been raised. The solid-state electrolytes are also weak against H_2O. In this chapter, solid-state electrolytes are categorized, and based on the weak point of such solid-state electrolyte, the treatments to overcome their weak points such as coating and doping with other metal ions are applied and then tried to handle them under atmosphere.

Keywords Solid-state electrolyte · All-solid-state battery · Oxide-based Li^+-ion conductor · Sulfide-based Li^+ ion conductor · Hydrogen sulfide gas · Doping · Coating

7.1 Solid-State Electrolytes and All-Solid-State Batteries

Recently, owing to safety issues, geometric restrictions related to electrolytes and inhibition of the formation of Li dendrites in LIBs, solid-state electrolytes (SSEs) have attracted much attention from researchers and engineers in the areas of batteries and energy. Various inorganic SSEs with Li^+-ion conductivities ($10^{-4} \sim 10^{-2}$ S cm^{-1} at 25 °C), electrochemical stabilities against lithium metal and wide electrochemical windows have been reported [1–3]. The possibility of the realization of high-power all-solid-state batteries (ASSBs) until 2030 has been raised [4, 5]. Polymer SSEs with Li^+-ion conductivity ($\sim 10^{-4}$ S cm^{-1} at 25 °C) have also been researched for a long time [6–8]. However, problems persist with the low cost of raw materials and the development of industrially scalable methods for the synthesis of SSEs and the preparation of ASSB [9, 10]. In addition, improvement of the stability of SSEs from the viewpoints of chemistry and electrochemistry should be examined in the bulk material of SSEs and at the interface between anodes/cathodes and SSEs [11, 12]. In particular, the stability of SSEs against H_2O is important for the development of ASSBs because it determines their fabrication process, performance and safety. As

© The Author(s), under exclusive license to Springer Nature Singapore Pte Ltd. 2022   57
F. Matsumoto and T. Gunji, *Water in Lithium-Ion Batteries*, SpringerBriefs in Energy,
https://doi.org/10.1007/978-981-16-8786-0_7

a result, an increase in the price of ASSB, which is an obstruction to the rapid spread of ASSBs, arises.

7.2 Inorganic Solid-State Electrolytes

Inorganic SSEs are classified into two categories of oxide- and sulfide-based Li^+ ion conductors. Moreover, the category of oxide-based Li^+ ion conductors comprises LiPON ($Li_{2.88}PO_{3.73}N_{0.14}$), perovskite ($ABO_3$, A: Li, La, Sr or Ca, B: Al or Ti), antiperovskite (Li_3OX, X:Cl, Br, I or mixture of halides), garnet ($A_3B_2(MO_4)_3$, A: Ca, Mg or Fe, B: Al, Cr, Fe or Ga, M: Si or Ge), NASICON ($AM_2(PO_4)_3$, A: Li or Na, M: He, Ti or Zr) and LISICON. The category of sulfide-based Li^+ ion conductors comprises thiol-LISICON ($Li_xM_{1-y}M'_yS_4$, M: Si or Ge, M': P, Al, Zn, Ga or Sb) and amorphous sulfide ($Li_2S-M_xS_y$, M: Al, Si, P et al.) [1, 3, 5, 11]. Some SSEs in sulfides react with moisture and generate poisonous hydrogen sulfide (H_2S) gas with the decomposition of SSEs [11]. Regarding H_2S generation in sulfides, Yu and Li published a review paper on the problems and methods for improvement [11]. The authors summarized a comprehensive evaluation of the chemical stability of SSEs against moisture (Table 7.1).

In oxide SSEs, the surfaces of SSEs react with moisture through H^+/Li^+ exchange reactions, as mentioned above. For example, the increase in pH value was obviously the consequence of the ion-exchange reaction of H^+/Li^+ in the SSE of

Table 7.1 Evaluation of stability issues of different electrolytes[a]

Solid electrolytes	Thermal stability	Chemical stability	Stability against Li	Electrochemical stability	Mechanical stability	Ionic conductivity
Polymer	**	***	***	**	**	***
Garnet	*****	**	***	***	*	***
NASICON	*****	****	*	****	*	***
Sulfide	***	*	***	*	**	****
Liquid electrolyte	*	*	**	***	****	*****

Reprinted with permission from Ref. [11]. Copyright 2020 American Chemical Society

[a]"Chemical stability" indicates the stability of SSEs against moisture. "Stability against Li" indicates the chemical and electrochemical stability versus the lithium metal anode. "Electrochemical stability" indicates the upper voltage limits of the SSEs. One star indicates a critical challenge of the SSE toward practical applications. Three stars suggest that the stability is sufficient to support the solid-state cell operation, but further improvements are still needed. For example, despite the garnet-type SSE exhibiting thermodynamic stability with lithium, interfacial modifications or some treatments (high temperature and pressure) are still needed to achieve a reasonable cycle stability. Five stars indicate that the stability of the SSEs is rather high and superior to other electrolytes

* Unacceptable

*** Sufficient for a working battery

***** Good performance

$Li_{6.6}La_3Zr_{1.6}Ta_{0.4}O_{12}$, leading to the formation of LiOH (Eq. 7.1).

$$Li_{6.6}La_3Zr_{1.6}Ta_{0.4}O_{12}(s) + xH_2O(aq.) \rightarrow Li_{6.6-x}H_xLa_3Zr_{1.6}Ta_{0.4}O_{12}(s) + xLiOH(aq.)$$
(7.1)

Thangadurai et al. also reviewed the chemical stability of garnet-type SSEs against moisture/humidity and aqueous electrolytes [14]. Because of stability in the context of moisture/humidity, although sulfide SSEs are unacceptable as SSEs for commercially available LIBs, as mentioned above, NASICON-type SSEs are highly rated among SSEs.

7.3 Polymer Solid-State Electrolytes

Conversely, polymer SSEs show the opposite trend; that is, H_2O contributes to improving battery performance. H_2O might be contained in the preparation process of polymer SSEs. For example, lithium salts with which base polymers are doped introduce H_2O into polymer SSEs. H_2O contained in SSEs contributes to the enhancement of the mobility of Li^+ ions in SSEs based on two mechanisms. The first is the action of H_2O as a plasticizer to improve the mobility of the polymer chains. The second is the higher affinity of Li^+ ions to H_2O than of polymer side chains. Li^+ ions that are not restrained by polymer side chains can be more mobile in SSEs, as shown in Fig. 7.1 [13]. Dollé et al. reviewed the impact of absorbed H_2O on the performance of SSEs via two mechanisms [13]. The question of how much H_2O degrades or improves Li^+ mobility in inorganic SSEs is of great interest. The data obtained with several inorganic SSEs are summarized in Table 7.2 [15–23].

7.4 Protection of Solid-State Electrolytes Against Humidity

Aguadero et al. analyzed the properties of $Li_{6.55}Ga_{0.15}La_3Zr_2O_{12}$ ($Ga_{0.15}$-LLZO) after H_2O immersion of $Ga_{0.15}$-LLZO at 100 °C with depth-resolved secondary ion mass spectrometry (Fig. 7.2) [24]. After H_2O immersion of $Ga_{0.15}$-LLZO at 100 °C for 30 min, the thicknesses of the corrosion layer of LiOH and protonated $Ga_{0.15}$-LLZO, resulting in H^+/Li^+ exchange reactions, were 0.15 and 1.2 µm. The bulk ionic conductivity of protonated $Ga_{0.15}$-LLZO decreased by 1 order of magnitude when compared with pristine $Ga_{0.15}$-LLZO. $Li_{6.5}La_3Zr_{1.5}Ta_{0.5}O_{12}$, which was exposed to humid air (humidity ~ 80%) for 6 weeks, formed LiOH layers with a 25 µm thickness, and the ionic conductivity of $Li_{6.5}La_3Zr_{1.5}Ta_{0.5}O_{12}$ became halved after exposure to the humid air [25]. Hwang et al. also reported dual-doped cubic garnet SSEs of $Li_{6.05}La_3Ga_{0.3}Zr_{1.95}Nb_{0.05}O_{12}$ with superior air stability and discussed their structural synergy for improving the air stability of the SSEs [26]. (Ca, Nb) Dual-doped

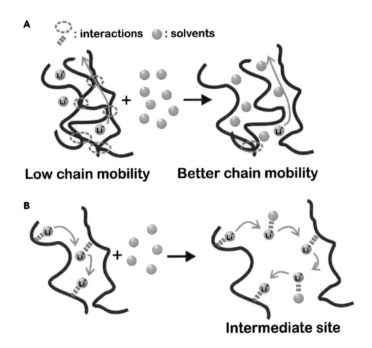

Fig. 7.1 Structural changes caused by solvent absorption impact lithium-ion mobility in polymer electrolytes. **a** Ion mobility in a dry polymer can be impeded by interactions between polymer chains, and residual solvent can act as a plasticizer and improve the mobility of the polymer chains. **b** Ion mobility is increased in the presence of solvent, as lithium cations have a greater affinity for polar solvents than for polymer side chains [13]. Copyright 2020, Creative Commons

LLZO SSEs exhibited an extensive degradation of charging/discharging cycle performance after exposure to air for 7 days and subsequent drying at 80 °C for 12 h, while (Ga, Nb) dual-doped LLZO SSEs exhibited stable charge/discharge capacities even after 50 cycles. From the Rietveld refinement of XRD patterns, dopant Ga preferably occupied the Li site and Nb the Zr site, while dopant Ca mainly substituted La in (Ca, Nb) dual-doped LLZO SSEs, which exhibited extensive degradation after air exposure. Doping in the Li and Zr sites of LLZO SSEs is a key point for improving air stability. The H^+/Li^+ ion exchange and structural stability of the high ionic conductivity Nb-doped Zr-garnet $Li_{6.75}La_3Nb_{0.25}Zr_{1.75}O_{12}$ (LLNZO) were investigated. In the study, the cubic garnet phase was confirmed to be maintained even though the exchange level reached up to 74.8%. This value indicated the replacement of octahedrally coordinated Li^+ in LLNZO. In addition, reverse ion exchange of H^+ by Li^+ was successfully achieved to be a reverse exchange of 75.3% at high Li^+ ion concentrations or strong basic aqueous electrolytes at 60 °C [27]. To protect SSEs from H_2O attack, the application of surface coating to SSEs was considered. Atomic layer deposition of Al_2O_3 on $Li_7La_{2.75}Ca_{0.25}Zr_{1.75}Nb_{0.25}O_{12}$ [28], chemical formation of dopamine-based shells on $Li_{6.4}La_3Zr_{1.4}Ta_{0.6}O_{12}$ (Fig. 7.3) [29] and LiF-coated $Li_{6.4}La_3Zr_{1.4}Ta_{0.6}O_{12}$ by NH_4F treatment [30] have been reported.

Table 7.2 Summary of the performance of solid-state electrolytes upon containing H_2O in solid-state electrolytes

Solid-state electrolytes	Information on the water content (relative humidity under storage, temperature, storage period, preparation condition of SSEs, water content)	Lithium-ion conductivity or capacity fading in charging/discharging cycles	References
LiPON	Pt/LiPON/Al test samples were left in ambient air for a period of 24 h with a relative humidity of 50% and a temperature of 22 °C	Ionic conductivity of LiPON 2.8×10^{-6} S cm^{-1} (before exposing to moisture) 9.9×10^{-10} S cm^{-1} (after exposing to moisture)	[15]
$Li_{0.38}Sr_{0.44}Ta_{0.7}Hf_{0.3}O_{2.95}F_{0.05}$ (LSTHF)	LSTHF pellets were immersed in water with different pH = 3, 7, 14 for 2 weeks	Ionic conductivity of LSTHF (25 °C) 4.8×10^{-4} S cm^{-1} (before exposing to water) 3.6×10^{-4} to 2×10^{-4} S cm^{-1} (after exposing to water)	[16]
$Li_{5+x}Ba_{x}La_{3-x}M_{2}O_{12}$ (M = Nb, Ta) (x = 0, 0.5, 1)	The 30 g was placed in a flask containing 200 mL of water and then allowed to spin using a magnetic stir bar and stir plate for 4 days at room temperature. The powders were then collected and dried at 700 °C for 2 h	The extent of H^{+}/Li^{+} ion-exchange $Li_5La_3Nb_2O_{12}$ 89% $Li_{5.5}Ba_{0.5}La_{2.5}Nb_2O_{12}$ 46% $Li_6BaLa_2Nb_2O_{12}$ 20% $Li_5La_3Ta_2O_{12}$ 64% $Li_{5.5}Ba_{0.5}La_{2.5}Ta_2O_{12}$ 45% $Li_6BaLa_2Ta_2O_{12}$ 20%	[17]
$Li_{6.6}La_3Zr_{1.6}Ta_{0.4}O_{12}$ (LLZTa)	The H^{+}/Li^{+} exchange was conducted via immersion of a fixed mass ratio (1:20) of LLZTa pellet in deionized water for 1, 3, 5 and 7 days at room temperature	The extent of H^{+}/Li^{+} ion-exchange 0 day 1.09% 1 day 44.19% 3 day 47.13% 5 day 47.93% 7 day 53.35%	[18]

(continued)

Table 7.2 (continued)

Solid-state electrolytes	Information on the water content (relative humidity under storage, temperature, storage period, preparation condition of SSEs, water content)	Lithium-ion conductivity or capacity fading in charging/discharging cycles	References
$Li_7La_3Zr_2O_{12}$ (LLZO)	LLZO was exposed to laboratory ambient air for 1 week	Ionic conductivity of LLZO (25 °C) 2.4×10^{-4} S cm^{-1} (before exposing to moisture) 1.6×10^{-4} S cm^{-1} (after exposing to moisture)	[19]
$Li_7La_3Zr_2O_{12}$ (LLZ)	The ion-exchange behavior of LLZ in water was analyzed by immersing powder samples (80 mg) in distilled water (100 mL) that was previously bubbled with N_2 gas for 24 h	Ionic conductivity of LLZ (25 °C) 2.8×10^{-4} S cm^{-1} (before exposing to water) 2.1×10^{-4} S cm^{-1} (after exposing to water)	[20]
$Li_{6.5}La_3Zr_{1.5}Ta_{0.5}O_{12}$ (LLZT)	The prepared samples were crushed into powder for the ion-exchange reaction. The ion-exchange reaction was performed by placing 0.5 g of the garnet oxides in a flask containing 100 mL water at pH = 7. The experiments were conducted under constant stirring at room temperature. The duration of the exchange process was 48 h	Ionic conductivity of LLZT (25 °C) 3.2×10^{-4} S cm^{-1} (before exposing to water) 3.2×10^{-5} S cm^{-1} (after exposing to water)	[21]
$Li_{1.3}Al_{0.3}Ti_{1.7}(PO_4)_3$	Pellet of $Li_{1.3}Al_{0.3}Ti_{1.7}(PO_4)_3$ was soaked in an individual pristine plastic vial with 3 mL of deionized water (with a resistance of approximately 20 MΩ cm) for the following periods of time: 10 min, 1, 2, 8, and 12 h	Ionic conductivity fades drastically after the water exposure. The conductivity losses after 12 h of soaking are as follows: 64% for total and grain boundary conductivities; 62% for bulk conductivity	[22]

(continued)

Table 7.2 (continued)

Solid-state electrolytes	Information on the water content (relative humidity under storage, temperature, storage period, preparation condition of SSEs, water content)	Lithium-ion conductivity or capacity fading in charging/discharging cycles	References
$Li_{1.5}Al_{0.5}Ti_{1.5}(PO_4)_3$ (LATP)	LATP were stored in ambient air for 72 h ($\sim$20 °C, relative humidity (RH) $\sim$50%)	Ionic conductivity of LATP (25 °C) 5.3×10^{-4} S cm^{-1} (after storing in ambient dry Ar) 1.0×10^{-3} S cm^{-1}(after storing in ambient air)	[23]

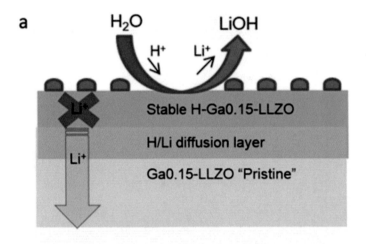

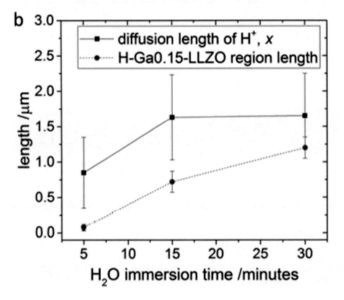

Fig. 7.2 a Schematic of the surface composition of the garnet grain and evolution with depth. **b** Plot showing the length of the H-$Ga_{0.15}$-LLZO region in the depth profiles (red dotted line) and the chemical diffusion length of H^+ (black full line). Reprinted with permission from Ref. [24]. Copyright 2018 American Chemical Society

ASSBs must possess waterproof properties because SSEs and Li metal are vulnerable to moisture in the atmosphere. The water vapor transmission rate of thin-film encapsulation of ASSBs should be investigated. Choi et al. proposed a hybrid thin-film encapsulation structure of hybrid SiO_y/SiN_xO_y/a-SiN_x:H/parylene and estimated the water vapor transmission rate of hybrid thin-film encapsulation to be 4.9×10^{-3} g m^{-2} day^{-1}, a value that is applicable to batteries [31].

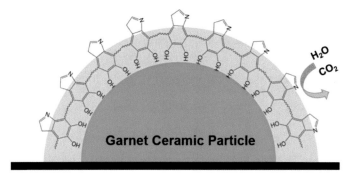

Fig. 7.3 Schematic diagram of air-stable DA-treated garnet ceramic particles [29]. Reprinted from J. Power Sources, 486, Jia, M.; Bi, Z.; Shi, C.; Zhao, N.; Guo, X.: Air-stable dopamine-treated garnet ceramic particles for high-performance composite electrolytes, 229363, Copyright (2021), with permission from Elsevier

Finally, as mentioned above, garnet SSEs are promising candidates for solid electrolytes for highly safe ASSBs; however, the SSEs are highly reactive to H_2O. $Li_{6.4}La_3Zr_{1.4}Ta_{0.6}O_{12}$ and $Li_{6.4}La_3Zr_{1.4}Ta_{0.6}O_{12}/MgO$ composite SSEs are generated by attrition milling, followed by a spray-drying process utilizing water-based slurries to form secondary granulates and to sinter green pellets prepared with secondary granulates. LiOH is dissolved in H_2O to prepare slurries to suppress the ion-exchange reaction between $Li_{6.4}La_3Zr_{1.4}Ta_{0.6}O_{12}$ and H_2O [32].

References

1. Gao Z, Sun H, Fu L, Ye F, Zhang Y, Luo W, Huang Y (2018) Adv Mater 30:1705702
2. Sakuda A (2018) J Ceram Soc Jpn 126:675
3. Zhang Z, Shao Y, Lotsch B, Hu YS, Li H, Janek J, Nazar LF, Nan CW, Maier J, Armand M, Chen L (2018) Energy Environ Sci 11:1945
4. Park KH, Bai Q, Kim DH, Oh DY, Zhu Y, Mo Y, Jung YS (2018) Adv Energy Mater 8:1800035
5. Pervez SA, Cambaz MA, Thangadurai V, Fichtner M (2019) ACS Appl Mater Interfaces 11:22029
6. Zhou D, Shanmukaraj D, Tkacheva A, Armand M, Wang G (2019) Chem 5:2326
7. Lopez J, Mackanic DG, Cui Y, Bao ZN (2019) Nat Rev Mater 4:312
8. Forsyth M, Porcarelli L, Wang X, Goujon N, Mecerreyes D (2019) Acc Chem Res 52:686
9. Hao F, Han FD, Liang YL, Wang CS, Yao Y (2018) MRS Bull 43:775
10. Sun YY, Li F, Hou PY (2021) J Mater Chem A 9:9481
11. Chen R, Li Q, Yu X, Chen L, Li H (2020) Chem Rev 120:6820
12. Zou Z, Li Y, Lu Z, Wang D, Cui Y, Guo B, Li Y, Liang X, Feng J, Li H, Nan CW, Armand M, Chen L, Xu K, Shi S (2020) Chem Rev 120:4169
13. Foran G, Mankovsky D, Verdier N, Lepage D, Prébé A, Aymé-Perrot D, Dollé M (2020) iScience 23: 101597
14. Hofstetter K, Samson AJ, Narayanan S, Thangadurai V (2018) J Power Sources 390:297
15. Nimisha CS, Rao GM, Munichandraiah N, Natarajan G, Cameron DC (2011) Solid State Ionics 185:47

16. Li YT, Xu HH, Chien PH, Wu N, Xin S, Xue LG, Park K, Hu YY, Goodenough JB (2018) Angew Chem Int Ed 57:8587
17. Truong L, Colter J, Thangadurai V (2013) Solid State Ionics 247–248:1–7
18. Yow ZF, Oh YL, Gu W, Rao RP, Adams S (2016) Solid State Ionics 292:122
19. Jin Y, McGinn PJ (2013) J Power Sources 239:326
20. Nemori H, Matsuda Y, Mitsuoka S, Matsui M, Yamamoto O, Takeda Y, Imanishi N (2015) Solid State Ionics 282:7
21. Li Y, Han J-T, Vogel SC, Wang C-A (2015) The reaction of $Li_{6.5}La_3Zr_{1.5}Ta_{0.5}O_{12}$ with water. Solid State Ionics 269:57–61
22. Pogosova MA, Krasnikova IV, Sanin AO, Lipovskikh SA, Eliseev AA, Sergeev AV, Stevenson KJ (2020) Chem Mater 32:3723
23. Dashjav E, Ma Q, Xu Q, Tsai CL, Giarola M, Mariotto G, Tietz F (2018) Solid State Ionics 321:83
24. Brugge RH, Hekselman AKO, Cavallaro A, Pesci FM, Chater RJ, Kilner JA, Aguadero A (2018) Chem Mater 30:3704
25. Xia W, Xu B, Duan H, Tang X, Guo Y, Kang H, Li H, Liu H (2017) J Am Ceram Soc 100:2832
26. Abrha LH, Hagos TT, Nikodimos Y, Bezabh HK, Berhe GB, Hagos TM, Huang CJ, Tegegne WA, Jiang SK, Weldeyohannes HH, Wu SH, Su WN, Hwang BJ (2020) ACS Appl Mater Interfaces 12:25709
27. Liu C, Rui K, Shen C, Badding ME, Zhang G, Wen Z (2015) J Power Sources 282:286
28. Hofstetter K, Samson AJ, Dai J, Gritton JE, Hu L, Wachsman ED, Thangadurai V (2019) J Electrochem Soc 166:A1844
29. Jia M, Bi Z, Shi C, Zhao N, Guo X (2021) J Power Sources 486: 229363
30. Duan H, Chen WP, Fan M, Wang WP, Yu L, Tan SJ, Chen X, Zhang Q, Xin S, Wan LJ, Guo YG (2020) Angew Chem Int Ed 59:12069
31. Lee HS, Lim KY, Kim KB, Yu JW, Choi WK, Choi JW (2020) ACS Appl Mater Interfaces 12:11504
32. Huang X, Lu Y, Jin J, Gu S, Xiu T, Song Z, Badding ME, Wen Z (2018) ACS Appl Mater Interfaces 10:17147

Chapter 8
Conclusion and Perspectives

In this review book, the advantages and disadvantages of H_2O in LIBs are reviewed, and the details of the reaction of battery materials such as cathodes, anodes and electrolytes with H_2O are summarized. Generally, considering factors of H_2O in the preparation process and battery performance, a water concentration of 100 ppm might be appropriate in conventional LIBs using organic electrolytes. Due to the waterproof property of cathodes and anodes and the addition of H_2O scavengers, the allowable range of H_2O will increase in the future to simplify the production process of LIBs. Conversely, at the level of properties of inorganic solid-state electrolytes, inorganic solid-state electrolytes that have recently attracted much attention cannot be in contact with H_2O. Although the development of technologies for leveling up the performance of ASSBs is a top-priority issue, the next step should consist of the development of technologies that endow waterproof property to inorganic SSEs.

Many researchers and engineers worldwide have expended much effort to study the development of post-LIBs, such as sodium (Na)-ion and Na metal, magnesium (Mg)-ion and Mg metal, calcium (Ca)-ion and Ca metal and Al-ion and Al metal batteries, which can overcome the weaknesses of LIBs, such as low battery performance, high cost, low natural Li abundance and battery safety issues. Many problems remained to be resolved for these batteries before their commercialization. Numerous materials and technologies have been examined and developed. The problems with the contact of these materials with H_2O were not discussed in this book due to limited space. The anodes, cathodes and electrolytes for post-LIBs exhibit the same issues as those in LIBs because the post-LIBs are operated under the same system as LIBs. Therefore, the ideas and materials developed for LIBs against H_2O might be applied to post-LIBs. However, fundamental issues in which cathode materials and solid-state electrolytes show weak effects against H_2O remain unsolved. Considering the points mentioned in this book, we think that the development of RABs with large potential windows due to water-in-salt electrolytes is an ideal goal. Despite the stability and charging/discharging performance of anode and cathode materials, the stability of anode and cathode materials will be a key point for the practical application of RABs with water-in-salt electrolytes.

F. Matsumoto and T. Gunji, *Water in Lithium-Ion Batteries*, SpringerBriefs in Energy, https://doi.org/10.1007/978-981-16-8786-0_8

Printed in the United States
by Baker & Taylor Publisher Services